Preface

READERSHIP

This book is intended primarily for students following a taught course in systems analysis.

  As a consultant in industry I have too often had the experience of teaching the principles of systems analysis to people who have been systems analysts for some time. The shortage of analysts, the urgency of business problems and the eagerness of many programmers and others to rise to new challenges have together conspired to press many a man or woman into a position of responsibility for which they were ill-prepared. Often, the results are very successful nonetheless; a testimony to the resourcefulness of the human in the face of practical problems and, sometimes, a testimony to many hours of midnight oil, unpaid overtime, weekend working, anxiety, missed holidays and neglected families. Sometimes the results are not so successful and the fruit of the analyst's labour lies abandoned, or is the cause of discontent, or requires major surgery at further cost. There is much in this book which could be useful for practising systems analysts who would like to increase their chance of getting it right next time.

SCOPE

The aim has been to produce a digestible text, tutorial in style, concentrating on essentials. The subject is a wide one and can be taken to include a vast amount of knowledge, technique, practice, folklore and theory. Three 'razors' have been applied to pare down the material: a razor of management, a razor of previous knowledge and a razor of payoff relevance.

Razor of management   Project management, management of data processing departments, business planning for data processing and the conduct of feasibility studies receive only skimpy treatment. The first and last are perhaps the most contentious exclusions, since they are nearly always among the duties of senior systems analysts. My motives are purely tutorial. These topics can be taught more effectively after the student is familiar with the processes being managed or planned. All the omitted topics are presented in a companion volume, Systems Management, which is suitable for reading after this book.

Razor of previous knowledge   Familiarity with computer hardware and software, programming from specification to testing, input-output devices, storage media and file organisation and access methods is assumed.

Razor of payoff relevance   I have asked myself, 'Is this information, technique or skill relevant to what most analysts have to do? Does this have a high probability of being of practical benefit to a trainee analyst?' This razor, more than the others, has dictated the content and balance of the book within its target size.

  Positive advice – things the analyst ought to do, how he can organise his thoughts or work – has consciously been preferred to negative advice – things which should not be done, pitfalls to avoid. This is partly because being positive is more enjoyable and partly because negative advice, once admitted, is rather limitless.

With the same intent to be positive, there is little contrast drawn between the practices suggested here and those in common use or recommended in other books. For those who want such contrasts, I would here draw attention to the early emphasis on and analytical approach to user participation; the consideration given throughout to specific participatory actions on the part of the analyst which will encourage the user towards self-help; a model of system development which does not over-simplify project phases; concern for requirements determination; the belief that data modelling is a tool which can help the analyst with any system, not just those destined for a Data Base Management System; the concern for human factors in system design and the belief that the human procedures and forms should be thoroughly checked out before programming; a methodology for file design and computer run design in which simplicity and flexibility tend to be preserved at the expense of efficiency in the first instance; a methodology which obliges the analyst consciously to review the important effects of his design rather than have some of them result by accident; the use of structured English as a program definition language; and, above all, a concern for the WHY as well as the HOW which will assist the analyst to adapt to new environments and overcome fresh problems.

ORGANISATION

The book is structured roughly as follows:

Explanatory text ⎫ repeat for ⎫
Questions, exercises ⎬ each section ⎬ repeat for
Discussion cases, assignments ⎭ ⎬ each
Answers to the questions ⎬ chapter
References ⎬
Appendices ⎭

The questions, and the answers to them, are an integral part of the text. Many of the questions aim to encourage discovery of important ideas. Subsequent sections often assume that the preceding questions have been tackled and the answers read. Some of the questions are simply reinforcers, encouraging rehearsal of the text or practice of technique. Others are very challenging and call for opinion, insight and intelligence. I have not dared to label the answers anything more than 'Answer pointers' – more thoughtful answers than mine will certainly be welcome. The reader is urged to develop his answer in his own original fashion, as he has much to gain from independent consideration of the question. He will be in a stronger position to criticise my answer or analyse the strengths and weaknesses of his own. He will develop a questioning approach which is a prized attribute of the systems analyst.

The discussion cases and assignments are reduced to a bare minimum and are meant only to illustrate some of the possibilities. A greater quantity of exercise and discussion can usefully be included in a taught course. A good source of ideas for further exercises and problems is **Information Systems: Theory and Practice** by J. G. Burch and F. R. Strater (Hamilton Publishing, Santa Barbara, California, 1974). J. Race's book **Case Studies in Systems Analysis** (Macmillan, 1979) contains cases suitable for more substantial practical assignments, syndicate exercises or class discussion. I shall be glad to supply further suggestions, cases, briefs for interviewing role-play, etc. to any interested teacher.

Systems Analysis

Andrew Parkin
Principal Lecturer in Systems Analysis
Leicester Polytechnic

Edward Arnold

First published 1980
by Edward Arnold (Publishers) Ltd,
41 Bedford Square, London WC1B 3DQ

Edward Arnold (Australia) Pty Ltd,
80 Waverley Road, Caulfield East,
Victoria 3145, Australia

Reprinted 1982, 1983, 1985

British Library Cataloguing in Publication Data

Parkin, Andrew
    Systems analysis.
    1. System analysis
    I. Title
    003   QA402

ISBN 0–7131–2800–3

Printed in Great Britain by
J. W. Arrowsmith Ltd, Bristol

REFERENCES

Books and journal articles by other authors have contributed at least as much to the ideas recorded here as my own direct experience. In keeping with the aims of concentrating on essentials, there has been ruthless selection of references for citation at the end of each chapter. Those from which an idea has consciously been stolen get referenced by way of acknowledgement. In addition, for each major topic, one modern source which itself cites many references has been included to aid the serious researcher. In óne or two cases, further citation of specially interesting work has been too tempting. Apologies are extended to all those authors and colleagues who recognise their phrases or ideas but who do not get the acknowledgement they deserve.

TERMINOLOGY

Those who like to fasten on to buzzwords and bandy them about will find lean pickings here. A new term might help to promote a new concept, but I am more concerned with the idea than the label. Sometimes I wonder if some of my computing colleagues, having discovered the right instantly to create a word of their own choosing when programming a computer, have too readily assumed a similar right when communicating with their fellows!
  It also has to be accepted that it is unrealistic to try to change accepted terminology, however good the case. A good example has been the widespread lobby to get the word 'informatics' accepted into English as a substitute for the irrational 'data processing'. Despite adoption of this word by many other European languages, 'informatics' is only limping along in academic circles in Britain and holds no sway with field practitioners. No doubt this is partly due to a cynical suspicion that it is a buzzword. No doubt it is also connected with the fact that rational PL/1 has not swept aside irrational COBOL, and for that matter the fact that we have not all thrown away our mother tongues and taken to Esperanto. People not only tolerate, but actually enjoy, idiosyncrasy and irrationality. They are likely to change their practices only if there is some good and urgent reason.
  In the original draft of this book, there was adopted (more by accident than by design) the phrase 'formal message system'. The idea of a formal message is very important to the subject of data analysis and I believe experienced systems analysts use the concept even though they may not articulate it. Clinging on to this phrase, I spoke of formal message systems which communicate those messages which have a predefined content. The phrase covered paperwork clerically prepared, as well as computerised systems. The term 'data processing' is often used loosely to include such applications as message switching or word processing where the messages are in natural language and not predefined. These are outside the scope of this book. Thus it seemed to me that 'formal message systems' preserved the generality I intended and was also more precise. However, taking the lesson of 'informatics', 'formal message systems' has been systematically replaced with 'data processing systems'.

COMPLETENESS

Nearly every chapter could have been the subject of a book in itself, but I did not set out to write an encyclopaedia or to cover every possibility. Although several sections have lists of points, the lists are not intended to be exhaustive of the possibilities nor comprehensive checklists for practical work. An attempt at completeness can be found in the book **System Development Methodology,** by G. F. Hice, W. S. Turner and L. F. Cashwell,

(North-Holland/American Elsevier, revised edition, 1978). The checklist approach adopted there will surely benefit the practising analyst who wants to be methodical in his consideration of alternatives and attention to detail.

ACKNOWLEDGEMENTS

I am particularly indebted to The National Computing Centre, Manchester, England, for granting permission to reproduce their standard forms from their publication Data Processing Documentation Standards. Likewise, I am grateful to The British Computer Society for permission to reproduce the society's Code of Practice and Code of Conduct.
 A large number of people have contributed specific ideas, criticisms and corrections. I would like publicly to thank Dr. K. D. Eason and Dr. D. R. Howe for their help.
 Anyone with a family who has tried writing a book will know that it is not only the author who bears the strain. My 'family' includes students who get a bit neglected when I have the bit between my teeth. My thanks to them for their forbearance. Most of all, though, I am grateful to my wife, Valerie, and children, Emma and Vicki, for their help and encouragement.

Leicester   A.P.
1979

Contents

1 Introduction to systems analysis

1.1 WHAT IS SYSTEMS ANALYSIS?

Systems analysis, for the purpose of this book, is defined by what a systems analyst does. A reasonable definition of what a systems analyst does could be as follows. In connection with a proposed computer-based data processing system, the systems analyst:

1 conducts a study of the feasibility of the system;
2 liaises with users of the system and determines their requirements;
3 finds out the facts important to the design of the proposed system;
4 determines the human and computer procedures that will make up the system, designing forms and files;
5 writes program specifications;
6 tests the programs and the system;
7 participates in the implementation of the new system;
8 documents the system.

One might add a practical note:

9 turns his hand to anything else within his competence which will further the organisation's desire for an effective and efficient system.

Systems analysis is an extremely difficult job to do well but it is very stimulating. There has been a consistent world-wide shortage of good systems analysts. Good systems analysts can earn very high salaries.

Questions

(Answer these as rationally as you can before looking at the answer pointers. The expected maximum time you should spend in thinking out and writing down the answers is given in brackets – you may choose to give up if you much exceed this. Answer pointers are given on page 6.)

1 'Teleology' is the name given to the practice of assigning human purpose to something which is non-human. For example, 'the roots of the plant seek out water' or 'the steam tries to equalise pressure by escaping through the valve'. Is there any phrase in section 1.1 above which strikes you as faintly teleological? (2 min)

2 An analyst/programmer is a person who is employed to do both systems analysis and programming. Some organisations have a policy of employing analyst/programmers as it makes for flexibility in allocating available staff to the required work. What effect do you think this policy has on (a) the quality of the analysis done? (b) the quality of the programming done? (c) staff recruitment and training? (5 min)

3 Which job is the most satisfying? (a) Systems analyst; (b) analyst/ programmer; (c) programmer. (2 min)

4 An orthogonal list of classifications is one in which the items on the list are mutually exclusive, the boundaries between items being clear, without gaps, so that the list covers the complete subject. What deficiencies are there in the orthogonality of the list of 'what a systems analyst does'

above?   (5 min)

1.2 THE PARTICIPANTS IN SYSTEM DEVELOPMENT

Overall responsibility for running an organisation is in the hands of 'top management'. This term might include a managing director and other working directors, executive officers, and executive committees. Of course, top management are answerable to boards, councils, courts, senates or to other groups representing shareholders, electors, taxpayers, and so on. Top management could be defined as the managers who are responsible for reaching agreement on plans with other power-holders of the organisation and for directing the everyday actions of the people within the organisation in such a way that these plans are carried out. In all but the smallest organisations this usually entails dividing the work among departments, each of which has a manager responsible for a function or objective defined by top management, e.g. sales, production, research and development, accounting, data processing. Large departments may be further subdivided into smaller sections or units. The departmental and unit managers usually have authority which is circumscribed by their superiors: they must do this, they may not do that, they must refer certain decisions for higher authority but can decide for themselves on others.
 The activities of the data processing department nearly always have, or potentially have, sufficient importance for the organisation to warrant the attention of top management. Even when data processing costs are a small part of an organisation's total costs, the consequences for the organisation may be great. Top management should be concerned with:

  getting the maximum benefit from data processing by ensuring that
    worthwhile opportunities are identified and successfully taken up;
  assessing the sensitivity of the organisational achievements to break-
    down, error or other disruption of data processing operations, taking
    steps to ensure that threats are suitably contained.

 In organisations where departmental managers are allowed, or required, to make most of their decisions without referring to higher authority, top management should be concerned with:

  making statements of policy to guide departmental managers in their
    decisions;
  establishing some method which will optimise the way in which data
    processing resources are allocated among departments;
  establishing policies or rules which the manager of the data processing
    department should follow when dealing with competing service requests
    from departmental managers.

 Where departmental managers normally have less discretion in executing their duties, or where it is decided that the managers will not enjoy their usual freedom specifically in connection with new data processing projects, top management should be concerned with:

  identifying opportunities for computerisation;
  making plans for implementing the systems and assigning priorities to
    the systems;
  defining responsibilities for implementation of the systems and monitoring
    their development.

 Top management may decide to increase the participation in this decision-taking by delegating their powers to a committee comprising representatives of themselves and other interested parties, e.g. managers of the affected departments, the data processing department, staff whose work may be affected. Such a committee is usually known as a steering committee.

A department affected by a proposed system is often called (by systems analysts) a 'user department'. The manager of such a department is a 'user manager', and the people in the affected department(s) are called 'users'. This last is a dubious term for two reasons. First, the term may embrace several groups in the department with different interests at stake – for example, people who prepare and handle data which is input to the system, people who receive output from the system directing their actions, people who get information from the system and use it to support their decisions, people who do not use the system directly but are affected by the decisions it supports or by the procedures it entails. Secondly, people who use the system or are otherwise affected by it may be outside the department most affected – people in other departments, the external auditors, customers, suppliers or even the public at large. Systems analysts sometimes include the people outside the affected department in the term 'user', sometimes not: some cynics have suggested they should be called 'victims' instead! Even without cynicism, it can be seen that some of the people called users are in fact servants of the system. Sometimes the term 'user' is applied loosely to embrace top management as well: in fact, anyone outside the data processing department.

This term 'user' has too much currency for an alternative seriously to be advocated. However, it is important to identify what it means in context and to make sure it does not obscure something important. It could be misleading to group together such varied classes of people under the single term 'user'.

The remaining participants in systems development are the systems analysts, programmers and operators who are usually employed in a central data processing department. They are controlled by a data processing manager, who assigns them to work on a given proposed system.

Questions

1 Systems analysts have been called 'agents of change'. An Agent is a person who represents a Principal. The Agent is supposed to further the interest of the Principal, in as much as the Principal has given him explicit or implied authority to do so, and he is supposed to refer to the Principal for a decision on matters outside his authority. An Agent's position is unsatisfactory if he represents more than one Principal and they have conflicting interests. In relation to a particular proposed system, whose Agent is the systems analyst? In other words, who is the systems analyst's Principal, to whom he should refer for a final decision on, say, the details of the system requirements? (5 min)

2 Should the analyst have authority to
a) decide the colour of the new bills sent to customers?
b) decide the best layout of a new report to be sent to a number of managers?
c) decide which people in a department should record new data required by the system?
d) decide how much the people affected by the system should participate in the design of it?
e) decide what data is to be held on the computer's files? (5 min)

1.3 PARTICIPATION BY USERS

The development of a new system is normally carried out by a team of analysts, programmers and users, although sometimes there are no users. The team is managed by a project leader.

User participation in system development could be classified as:

a) 'insider participation', in which the participant is project leader or

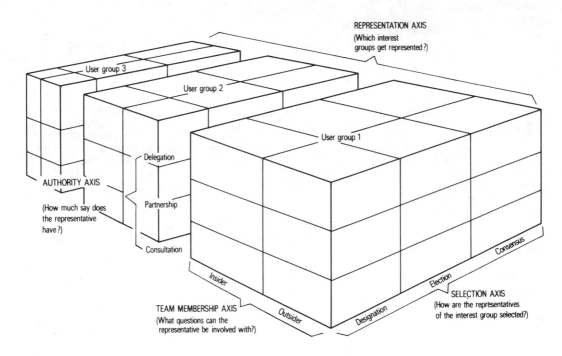

Fig. 1.1: Dimensions of participation

 another member of the project team;

b) 'outsider participation', in which the participant is not a member of the project team, but has some other opportunity to influence system development.

Insiders are usually privy to all design decisions, so they may have some opportunity to influence any aspect of the system design should they wish to do so.

 User participants may be **designated** by the sponsor or by a subordinate acting on the sponsor's authority (the sponsor might designate himself), they may be **elected** as a representative of a user group, or they may comprise an entire user group, a **consensus**.

 Outsider user participants may be involved through **consultation, partnership** or **delegation**. In consultation, the project team analyses the system and produces a planned design. Then the opinions or preferences of users are sought and these are taken into account as far as possible. Although consulted participants may have no authority over the system design, in practice it is usually difficult for a project team to ignore their opinions or preferences without a convincing reason. Sometimes a consulted participant may have power of veto.

 In partnership, the project team describe the problems they see to the users (and vice versa) and the two groups work together to produce a joint design of the new system or some aspect of it. In delegation, a user group is given some portion of the system to design for themselves, and they can exercise autonomous choice over the features of its design.

Insider participants are nearly always involved through partnership or delegation, although it is not impossible for an insider to have a purely consultative role.

If participation does not take place, users will probably be **told** to accept a new system designed by the project team (perhaps told indirectly, by being presented with a fait accompli) or they may be **sold** on the new design. Selling may take the form of persuasion to accept the new system, either by stressing the benefits to the users or by offering compensation so that the cooperation of the users is, in effect, bought.

Although these dimensions of participation have been presented in neat compartments, in practice they are more like points in a spectrum of possibilities (see Figure 1.1).

Questions

1   It has been said that user participation is desirable because it will lead to a more **effective** system. Do you think this is likely to be so? Can you think of arguments for and against?   (20 min)

2   It has been said that user participation is desirable because users have a **right** to influence the system design. Does such a right exist?
                                                                        (2 min)

3   Do you think users have a **responsibility** to participate in the system design?   (5 min)

DISCUSSION CASES

1   The management of a medium-sized factory were considering an improvement to their system for maintaining control over their stocks of raw materials. The factory had a friendly atmosphere and all the workers called the plant manager 'Dad'. It was clear from the outset that the new system would entail small changes to the way the production workers recorded their output at the end of the day, and more substantial changes to the procedure followed by the storeman who was in charge of raw materials. The systems analyst assigned to the job got on friendly terms with the production workers, the storeman and the production department managers. He involved all these parties through partnership, using the production foreman as the representative of the production workers.

When the analyst presented his description of the proposed system, about six months after starting, it contained no surprises to anyone and he was sure there would be no objections - or, if there were, they would be very minor points which could be easily fixed. It was his opinion that although the involvement procedure had been time-consuming and required more effort, this was more than offset by the increased effectiveness of the system and the acceptance of it by everyone concerned. The managers were completely satisfied with his efforts.

2   A systems analyst was assigned to implement a new payroll system in a plastics plant, during a period of high unemployment. The comptroller, who supervised the three clerks responsible for preparing the payslips for the plant workers, had already decided that the new system was to be a standard one run by a local computer service bureau. The purpose of the new system was mainly to reduce mistakes, to increase reliability and to improve information available to management. The existing system was overstretched.

When the systems analyst interviewed the payroll clerks, it was clear that they were very apprehensive about how the new system would affect their jobs. The systems analyst implored the comptroller to explain to them company policy about the new system. The analyst also wanted to consult the payroll clerks, or involve them as partners, with regard to the part of the system which affected them, but the comptroller declined, on the grounds that it was nothing to do with them. The system was implemented about two months after the analyst was assigned. Shortly before it became operational, two of the three payroll clerks resigned, mainly because of their apprehension over the new system and their resentment at what they considered the high-handed attitude of the comptroller. The comptroller was easily able to fill the vacancies. New clerks were trained and in office with virtually no disturbance to the system. The comptroller was very pleased with the system when it was operational.

ANSWER POINTERS

Section 1.1

1   If the term 'organisation' in item 9 of the list describes an organised system of people, plant, premises, materials, money, etc., then the phrase 'organisation's desire' is somewhat unhappy. Presumably it is only the people who have desires.

2   There is nothing in the rules to say that an analyst/programmer should not both analyse as well as a systems analyst and program as well as a programmer. There are people who would perform well in either capacity. Against this, systems analysis and programming are both demanding jobs. It takes an exceptional person to be a good programmer and an exceptional person to be a good analyst. It may be considered that it would take an even more exceptional person to stay good at both. Experience suggests that analyst/programmers are generally **either** good at analysis **or** good at programming – generally the latter. This may have more to do with the difficulty of maintaining training and practice across such a broad front than with innate personal characteristics.
  A reasonable answer to the question is therefore that the price paid for flexibility in allocating staff is **either** some reduction in the quality of the analysis **or** some reduction in the quality of the programming **or** increased difficulty in finding or training the desired calibre of personnel – or some combination of the foregoing.
  It would be wrong to leave the impression that the only justification for analyst/programmers would be increased ease of assigning people to jobs. It may also be a good way to train a programmer in systems analysis. It may allow programmers to take on some of the analysis work which is more concerned with the computer, leaving the systems analyst more free to concentrate on the business side of the system. It may provide increased satisfaction for a particular programmer who needs a more demanding job. It may help involve an analyst in the implementation of his end product.

3   If job satisfaction depends on the extent to which a job meets the job-holder's aspirations, expectations and needs, assertions about satisfaction can be made only in relation to a particular person or, possibly, a class of person with known characteristics; this question does not make sense since it implies that satisfaction is independent of the people concerned.

4   The more obvious ones are as follows: the 'requirements of the users' (item 2) are 'facts important to the design' (item 3), so this suggests some overlap; item 7 could contain all the others unless the word 'implementation' were narrowly defined; item 9 is phrased in catchnet fashion, picking up anything that slips through the others, so if it has any substance the others must be deficient in covering the complete subject. Item 9 is also somewhat tautological, since it is coming close to saying 'a systems analyst does ... anything else that a systems analyst does'. It may also be considered that 'conducting a study' (item 1) could entail 'liaising with users' (item 2)

and 'finding out facts important to the design' (item 3), perhaps more.
   There is not necessarily anything wrong with classifications that are not orthogonal (it all depends on the use to which they are put) but increased orthogonality often increases usefulness or understanding. Perfect orthogonality is hard to achieve; perhaps impossible when the thing being classified is a real-world system, as opposed to an abstract one. The concept of an ideal orthogonal division is important in systems analysis and will arise several times later in the book.

Section 1.2

1   The Principal, it could be proposed, should be someone with the authority to make decisions about matters which involve conflicting interests of different user groups. He probably authorised the systems analysis and design in the first place. Thus, where departments are autonomous, and where a departmental manager has commissioned a system, the department manager is the Principal. Otherwise, the Principal will be the top manager who authorised the work, or the steering committee if it exists and if it authorised the work. Although a steering committee is not an individual and comprises people whose interests may conflict, the end product of debate in the committee is that it speaks with a single voice to outsiders, so is acceptable as a Principal.
   I have in the past called this person/body the Commissioner and have also heard the term Prime Client used in the same connection. One author calls this person the User, which is rather confusing. At least we are agreed that he is an important participant who should be identified by the systems analyst, even if there is no widely accepted term. Possibly the most apt term, which is gaining currency, is 'Sponsor'; this is the term which will be used for the rest of this text.
   Perhaps it should be pointed out that just because someone has called systems analysts 'agents of change', it does not follow that they are and that they must have Principals. The idea of a systems analyst exercising personal authority over conflicts of interest, deciding what is best and answerable to no-one save his own conscience, has a certain whimsical appeal, like the Lone Ranger; but he is unlikely to find many organisations where this view of the power he should have is shared by others.

2   There is nothing to stop top management delegating authority to whoever they think fit, although an exception may occur with some responsibilities which are defined by law. If they believe that the systems analyst would be the best person to make decisions on the matters listed, they should give him the authority. In the absence of express authority, though, each of the decisions listed – colour of bills, report layout, personnel assignment, participation in decision-taking, what data is kept on record – will be the responsibility of some other person or persons in the organisation, probably the managers of the affected departments. Assuming that top management continue to have confidence in those other persons, the systems analyst should not have authority to make the decisions listed. So my answer to each part of the question is 'probably not'. However, this does not mean that the systems analyst may not have an opinion on these matters, nor that he should not volunteer advice to the decision-taker nor, indeed, that he cannot try to persuade the decision-taker if he believes a poor decision is being taken.

Section 1.3

1   Users who have not participated will be told or sold. A told or sold user is more likely to react negatively to a new system than one who has participated to the point where he is confident the system will satisfactorily cater for his needs and preferences. A negative reaction by a user may

take the form of hostility to the change, lack of cooperation, or diminished enthusiasm. If this results in less effectiveness, then lack of participation is likely to lead to reduced effectiveness. If the user has participated to the point where he identifies the new system as being the result of his own choice, then he is less likely to react negatively when it is in operation.

Users have expert local knowledge, i.e. knowledge of facts which are important to that part of the system with which they are concerned. They may because of this be capable of making a more effective design which caters for local conditions, exceptions, preferences and interactions, ensures that local group social needs are satisfied, takes advantage of local opportunities and avoids local risks or pitfalls. A systems analyst is unlikely to acquire equivalent knowledge within a reasonable time.

Any system of which men are a part depends for its effectiveness on the performance of those men. Systems analysts sometimes fall into the trap of supposing that the human part of the system will operate mechanistically in a predetermined manner. Particularly when the human job is routine or low-skilled work, there is a tendency to assume that 'anyone can understand how that job is done', or to assume that user reactions are 'obvious'. This is probably a mistake; humans are complex and it is very difficult to predict the reactions or performance of men and women in a complicated system such as a data processing system. Even a simple data processing system involving humans must be considered complicated compared with a mechanical system. In practice, the only reliable indicators of the acceptability, ease of use or satisfactoriness of the system in operation are likely to come from those who will be operating it. This is a strong argument for user participation, at least at the level of consultation about the aspects of the system which concern them, or for having the users take part in experimentation of the proposed system, so that their reactions can be established.

Users without suitable aptitude, training and experience, may be unable to design a system which is as technically effective as that designed by a systems analyst. They may make poor choices on matters that do not directly affect them – for example, the timeliness, accuracy, reliability or security of data passed outside their group. They may miss technical opportunities through lack of knowledge of possibilities. They may place wrong emphasis on the importance of different parts of the system with which they are concerned. They may design a system which is open to disruption or other risk in circumstances which are not presently prevailing but which might be foreseen by an analyst. They may fall into pitfalls familiar to the analyst.

The more extensive the representation, the more selection tends towards consensus and the more there is insider participation, so the greater are the difficulties of organising and coordinating participation, the more likely it is there will be conflict over how the system should operate, the more debate there is likely to be in attempts to reconcile conflicts and the longer the time scale for implementation, possibly resulting in lost benefits. The grand purpose of a new system may be lost if the design responsibility is distributed far and wide. If conflicts remain, after having been exposed and sharpened by debate among the representatives the eventual choice may cause more discontent, or the choice will be postponed for fear of the eventual consequences.

My answer to the question 'Is participation effective?' is 'Yes, usually, but exactly what sort of participation are you talking about?' The most effective form of participation will be one which fits the particular system, the particular people and the particular priorities of the moment.

2   The reason only two minutes were suggested for this question is that I thought either you would have strong views on this subject and would deliver an immediate opinion, or you come from a country like Norway or

West Germany where trade union representatives have a legal right to be consulted on proposed changes in methods of working, or you recognised this question as embracing an issue of our modern era, perhaps **the** issue in industrialised countries, i.e. to what extent should the employees of an organisation have rights in the management of the organisation. In the United Kingdom, there is no **statutory** right of participation, although there may be some **contractual** rights to consultation included in agreements between unions and employers. Perhaps more important for the present discussion is whether or not the employee has a moral right. An important corollary of a right is that there must be a **duty** or obligation on the part of others to respect it. So, to ask 'Does an employee have a right to influence system design?' is also to ask 'Does an employer have a duty to allow the employee to influence the design?'

Ethical questions like these can rarely be dealt with in terms of absolutes or objective facts, and differences of opinion are likely. Confining the discussion to the industrialised west, a reasonably unassailable observation is that there has been a shift of opinion. Perhaps a generation ago, or earlier, management was seen more as a private activity, with the manager accountable for his actions only to his superiors and the owners. Nowadays, management is more public, and managers are more often obliged, or choose, to account for their actions to employees or to society. My opinion (although my mind is by no means closed to argument) is that in most cases a manager should have the right to decide what work is to be done, the target quality of it and which employee or group of employees are to do it, assuming it is under the general terms of their employment. The quantity of the work and the resources applied to it should be bartered or agreed between the manager and the employee (with appeal to independent referees if necessary). The employees should have the right to decide how the defined quantity and quality of work should be done with the defined resources.

It is also my opinion that whatever rights exist on the part of other parties to influence the quantity or quality of the work should generally be exercised through the method of appointing the manager, exacting accountability from him and rewarding him. In the case of monopolies and public administration, I think customers and clients should have additional rights of consultation when a change of service is involved.

3   It has already been suggested that top management has responsibility to take advantage of opportunities and safeguard against risks, as do other sponsors. The users with the authority also have a responsibility to decide on the details of the system where relevant, e.g. colour of the bills, report layouts, personnel assignments, extent of participation and what data is to be recorded. These are specific responsibilities which exist in an organisation and for which the user is accountable to a superior. If authority to do the design is delegated to users, and they accept this authority, then they are also accepting responsibility and should expect to be accountable for their efforts.

In the case of responsibility to oneself, for example 'everyone has a responsibility to vote', that is a matter of personal choice.

REFERENCES

(1) Aron, J., **Information systems in perspective**, Computing Surveys, 1 4 pp. 213–36.
(2) Parkin, A., **The scope of systems analysis**, Computer Bulletin, No. 15, March 1978, pp. 10–11.
(3) Weir, M., **Managing the implementation of computer systems – a comparison of approaches**, in Parkin, A. (Ed.), **Computing and People**, Edward Arnold, 1977, pp. 7–12.

(4) Damodaran, L., **User involvement in system design – why? and how?**, in Parkin op. cit. above, pp.13–19.
(5) Boneham, P., **Industrial democracy and systems design,** Computer Bulletin, No. 20, June 1979, pp. 10–11.
(6) Mumford, E. and Pettigrew, A., **Implementing strategic decisions,** Longman, 1975.
(7) Mumford, E., and Henshall, D., **A participative approach to computer system design**, Associated Business Press, 1979. See also the correspondence in **Computing**, 30th August to 8th November 1979.

2 The system life cycle

2.1 THE SYSTEM LIFE CYCLE

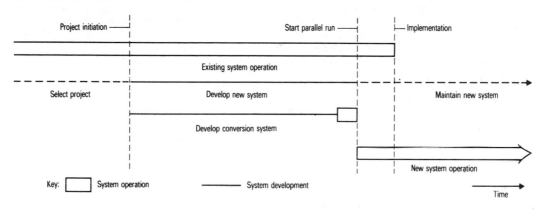

Fig. 2.1: The system life cycle

Figure 2.1 gives a simple model of how a new system may replace an old one. It is difficult to generalise about what is meant by a 'system' (to be more precise, such generalisations tend not to be helpful) because the meaning of the word in a particular case is defined by the scope of that project. Thus, if we speak of 'the invoicing system' we are speaking of those components and processes of invoicing which we **choose** to consider relevant or important. The scope of the invoicing system is not an objective fact which exists independently of people; the scope is defined subjectively. A system is often called an 'application' by systems analysts because it fulfils the business purpose to which the computer is **applied**.

Prior to project initiation, little formal work may have been done on the system, although the projected change may have existed in the minds of management for several years. The process of identifying which particular new system to develop, out of all the new systems that could be developed, is often called 'project selection' or 'application selection'. If these phrases suggest that some amassing and classifying of alternatives takes place, one alternative being chosen according to prescribed criteria, then they are not very descriptive of what often happens in practice. In practice, projects may be promoted by an event such as:

a) a difficulty being encountered in operating an existing system (perhaps the invoices are not coming out fast enough, or not enough people can be recruited to do the processing, or a lot of mistakes are being made;

b) a manager hearing of a successful system in another organisation and seeing a similar application in his own;

c) a manager becoming dissatisfied with the level of indirect expenses (administrative expenses such as those incurred in producing invoices), which are not directly concerned with supplying the goods or services of the organisation;

d)   a sales representative from a computer manufacturer or computer service bureau encouraging a manager to take an interest in a particular application.

One authority has aptly suggested that we should speak of projects being 'triggered off' rather than 'selected'.

After project initiation, work proceeds on the development of the new system which will replace the old. At the same time, a **conversion** system often has to be developed and brought into operation prior to introduction of the new system. The conversion system prepares the ground for changeover from the existing system to the new system. If the new invoicing system relies upon customers' names and addresses being recorded on magnetic tape files, and they are at present recorded by typing on index cards, then an objective of the conversion system could be to transfer customer names and addresses from the index cards and keep them up-to-date on magnetic tape until the new system is fully operational and can take over the updating process. (This example should not be taken as meaning that customer names and addresses are always converted in this way; see chapter 16.)

It is often arranged that the new system and the existing system operate concurrently for some period, as illustrated in Figure 2.1. This is to allow the accuracy of the new system to be established by comparing its results to those of the old. This period is known as 'parallel running'. After the new system becomes operational, some modifications are normally required, and 'system maintenance' is the term applied to the small improvements, changes and corrections which are made to the new system procedures.

Questions

1   If 'implementation' of the new system is regarded as a point in time, do you think that Figure 2.1 is properly labelled, or should the arrow of 'implementation' point to the same place as 'start parallel run'?   (5 min)

2   Draw a diagram similar to Figure 2.1 for each of the following cases.
a)   A system is to be implemented to provide weekly reports to managers about customer account balances. No equivalent system is in operation at present, and all the data for the new system is already held on computer files.
b)   A company has outgrown its present premises and is considering where to relocate. A computer system is to be developed which will aid this decision by reporting how different locations would affect the cost of distributing the company's goods. No such system exists at present in the company.   (5 min)

3   A **descriptive** model is any means of describing or explaining what **does happen** in a system. A **normative** model is one which seeks to explain what **should happen** if some target or objective is to be reached. 'People usually come into the park by the North Gate and go to the South Gate via the Eastern Bandstand' is descriptive, whereas 'If people wish to get to the South Gate as quickly as possible from the North Gate, then they should follow the flagstone path joining these two gates directly' is normative.
a)   Is the model of the system life cycle, illustrated in Figure 2.1, a descriptive model or a normative model?
b)   Is the list of project triggers in the text normative or descriptive?
                                                                      (2 min)

4   A **precedence network analysis** can be used to schedule or describe inter-dependent activities. Each activity is represented by an arrow preceded by and followed by a circle representing events. An activity therefore starts with a begin event (at the tail of the arrow) and terminates with an end event (at the head of the arrow). Getting a tailor-made suit could be described as in Figure 2.2. The level of detail or assumptions can be chosen

Fig. 2.2: Network - the tailor-made suit

by the analyst. If we wished to show the dependence of choosing the suit style and cloth, these could be added as in Figure 2.3. The dummy arrow shown by a dashed line is used to satisfy a convention that no two activities share the same head and tail event (this convention is desirable for uses of the network with which we are not at present concerned).

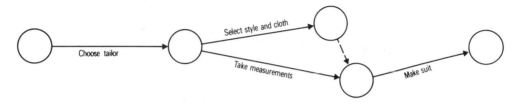

Fig. 2.3: Figure 2.2 revised

Maybe you are more likely to choose the style and cloth before measurements are taken. This does not affect the network, which is concerned only with the logical dependence because of physical constraints. Thus the network diagram in Figure 2.3 claims that taking measurements could precede or follow the choice of style and cloth, or the two operations could take place in parallel. However, a start cannot be made on making the suit until both those activities are completed. There is no implication of time scale in the length of the arrows. Labelling the events usually adds little to the meaning so event names are usually omitted.

Figure 2.4 gives a further example.

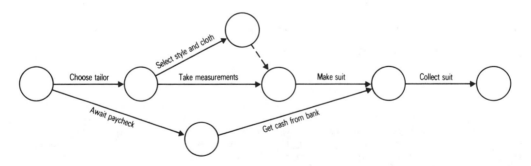

Fig. 2.4: Figure 2.3 revised

a) State three assumptions the analyst has made implicitly in the example network of Figure 2.4.
b) Construct a precedence network diagram labelling it with the following activities: select project; plan new and conversion systems; develop new system; develop conversion system; operate conversion system; parallel run; operate new system.   (10 min)

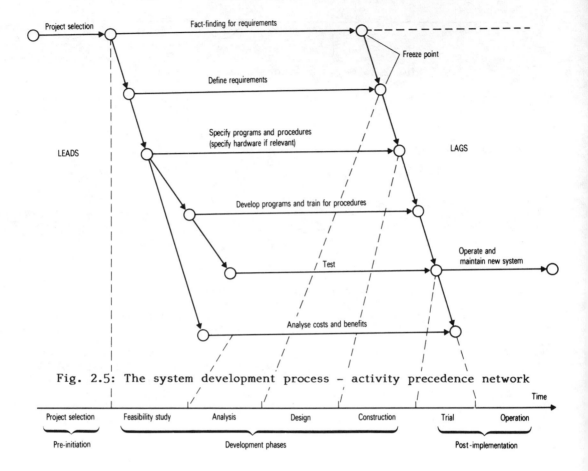

Fig. 2.5: The system development process – activity precedence network

Fig. 2.6: The system development process – project control phases

2.2 THE SYSTEM DEVELOPMENT PROCESS

The solid lines in Figure 2.5 above show a precedence network which looks somewhat like a ladder and which describes the dependences involved in developing a new system. The rungs of the ladder show the activities that best describe the process.

The LEAD activities making up the left side of the ladder are intended to show that a lower rung activity cannot start until some time after a higher rung activity has started. Definition of requirements cannot start until some fact-finding has been done – the first LEAD activity can be interpreted as 'get enough facts to begin definition of the requirements'. Specification of programs cannot start until some requirements and procedures have been defined. Reliable estimates of costs and benefits cannot be made until programs and procedures are known at least in outline. Programming and procedure training cannot start until at least one program or procedure is specified. Testing cannot start until one program or procedure is ready to be tested.

The LAG activities making up the right side of the ladder show that a lower rung activity cannot be completed until some time after the activity above has been completed. The network asserts that requirements cannot be finalised until enough relevant facts are known. Program specification cannot be completed until all requirements and procedures are determined. Programming and procedure training cannot be completed until all programs and procedures are specified. Testing cannot be completed until all programs and procedures are ready. Costs and benefits cannot finally be determined until after the project is completed and the new system is operational (if then!).

Returning to the fact-finding, 'enough relevant facts' was written rather than 'all relevant facts' because it is unlikely in practice that **all** relevant facts can be found. The relevant facts are not a stable and objective set of facts; they change as the design features and requirements of the system change. These in turn change with people's preferences or knowledge, and with pressures from the environment. To look for all relevant facts would be a never-ending task. This leads the pragmatic systems analyst to draw the line at the point where he judges that enough relevant facts have been gathered to enable agreement to be reached on the requirements of the system. This point is known as 'the freeze' and after this point the system requirements are considered 'frozen'.

Of course, this human attempt to crystallise the objectives is not mirrored in the real world. After the freeze, it is often the case that unsuspected facts prove relevant as the design is made in greater detail and as new problems are encountered as a result of the progress made towards completion of specifications, programming and so on. Also, the requirements themselves may change as new business possibilities or pressures emerge, or as new users replace the old. The declaration of the freeze does not preclude the need to note further relevant facts, and the dotted line continuing the top rung of Figure 2.5 illustrates this.

Although the flow of dependence is generally down the ladder, there is a feedback of information from lower rungs which may modify activity in a higher rung. For example, although analysis of costs and benefits cannot begin until after some requirements are known, preliminary analysis of costs and benefits may change expectations and may result in the definition of different requirements.

The dashed lines map the network model of Figure 2.5 onto the simpler straight-line model of Figure 2.6. The latter shows the project on a time-scale, split into phases which have been chosen, somewhat arbitrarily, for the purpose of exercising managerial control over the development of the new system. The end of each phase is punctuated by 'deliverables'. These are tangible evidence of achievements which can be reviewed by management to assess progress and to decide, in the light of expected costs and benefits, whether to continue, to abandon or to modify the project.

Often, an attempt is made to marshal the activities into the phases; this is such a common practice that many systems analysts identify activities as 'belonging' to a particular phase. In such cases, specification is not officially started until requirements have been agreed, and programming is not started until all program specifications have been written. This process is known as 'lock-step project control' and there are arguments to justify it when the opportunity costs of the analysts and programmers deployed are high.

It is difficult to generalise over the likely duration of a project or its phases. This is especially so with the feasibility study phase, at the end of which the sponsor has to decide whether or not to authorise the rest of the work, which is usually more substantial. He may decide to postpone this decision and to await developments or different priorities. If work on the project is delayed, the sponsor may later decide that feasibility should be reassessed and costs and benefits re-estimated. Sometimes several successive feasibility studies and reports are made; this process can take years.

If the balance of the system development is to take less than one or two months, it is likely to be classified as a system maintenance operation. If a project is expected to last much more than a year, a substantial outlay will probably be involved before any returns are seen. It would be less risky to identify some beneficial subsystems which can be implemented in a shorter time.

Typically, 10–20% of total system development financial costs are incurred in the feasibility study. About 40% are incurred in fact-finding, determining and specifying requirements, following the feasibility report; 40–50% in program development and testing.

Questions

1   The following are other names given to system development phases. Relate them to the most synonymous term in Figure 2.6. Why do you think are there so many different names?

Project request; system proposal; system specification; implementation; system audit; program and procedure specifications; initial study; requirements definition; definition study; post audit; program and human job development; testing; acceptance testing; detail design; system investigation; post-implementation evaluation; current systems research; proving; programming; information analysis; system analysis and design; preliminary survey; system implementation; outline design; installation; functional specifications.
(5 min)

2   What shoud be done about relevant facts or new or changed requirements which emerge after the freeze?   (10 min)

3   Sometimes it may be decided that the costs and benefits of a system cannot be determined with sufficient reliability until some of the new procedures or programs have been experienced and checked out. How would Figure 2.5 have to be amended to cater for this?   (2 min)

2.3 THE DELIVERABLES

The phases and deliverables suitable for a project of about 3–12 months duration, using a project team of 1–6 members, could be as in Figure 2.7. For a longer or more substantial project, it is likely to be desirable to specify additional phases or deliverables so that suitably detailed control can be exercised over project development.

Learning how to undertake an effective feasibility study, prepare a project plan and analyse costs and benefits is easier after other systems analysis functions are understood, so these topics are not covered here but in Systems Management, along with such topics as project selection, staffing and personnel management.

The objectives of a feasibility study could be summarised as follows:

a)   to consider alternative designs and reject the unacceptable ones for any good reason – economic, technical, organisational, social, time-scale;

b)   to choose the most promising remaining case and to develop the design of this to a point where a reliable prediction can be made of its costs and benefits and reliable assurance can be given that it does not hold unacceptable economic, technical, organisational, social or time-scale side-effects;

c)   to lay the foundations of subsequent development work should it be authorised.

The word organisational here particularly refers to the structure and long-range plans of the organisation; the word social particularly refers to the satisfaction of user groups affected by the system.

Phase	End of phase deliverables
Project selection	Terms of reference
Feasibility study	Feasibility report to the sponsor Project plan Analysis of costs and benefits
Analysis	Requirements report to the sponsor Revised project plan Revised analysis of costs and benefits
Design	Program specifications and database definition Training manuals Hardware specifications where relevant Revised project plan Revised analysis of costs and benefits
Construction	Documented programs Test data and results Supply of consumables The working system
Trial	Post-implementation evaluation report to the sponsor Analysis of costs and benefits

Fig. 2.7: End-of-phase deliverables

A feasibility report usually contains the following information:

Justifications: statement of system objectives;
                     alternatives considered and preferred solution;
                     cost and benefit summary;
                     limitations of preferred solution.
Present system: summary of procedures;
                     list of inputs, outputs and files;
                     control and security features.
Proposed system: summary of procedures;
                     outline design of inputs and outputs;
                     database contents;
                     control and security features.
Conversion: requirements and plan.
Appendices: all supporting detail including project plan and detailed
                     analysis of costs and benefits.

A requirements report should contain:

details of outputs, inputs and data to be stored;
detailed business procedures to be adopted in user departments;
outline computer procedures required;
appendices of supporting detail, including revised project plan and
   revised analysis of costs and benefits.

Questions

1   It has been suggested that an important criterion of project selection
is that the project's value should be readily recognised by user management
and the sponsor, even if this would mean that a more valuable – but less
obvious – opportunity is foregone. Do you agree with this? Explain.   (5 min)

2   Sometimes, after a system is implemented, a dispute arises between the
sponsor and the project team as to whether or not the new system performs
effectively. The dispute is often rooted in different assumptions or different
subjective judgements. How might such a dispute be avoided?   (10 min)

3   How would you describe the activities which should be undertaken in a feasibility study? Make a list of about ten items.   (20 min)

4   The phrase 'costs and benefits' has been used several times so far in this chapter. How would you define 'costs and benefits'?   (10 min)

DISCUSSION CASES

1   The General Manager of CABCO, a small machine tool factory in the United Kingdom and a subsidiary of a large US multinational, decided to apply a computer to keep accounts of the many suppliers and stocks of bought-in materials and parts, and to produce management reports. This was the first attempt to use a computer in the factory. The supplier and stock accounting was done by a chief clerk and two bookkeepers, but it was a growing system and the need for a further bookkeeper was likely. The company rules required that any expenditure on data processing must be authorised by the US parent company, which in turn meant that the general manager would have to make out a case in justification.

  With the aid of sales representatives of two computer manufacturers who had been invited to make recommendations, the general manager and the chief clerk worked out quite a detailed description of what the computer would do. They also calculated that the routine work of one of the book-keepers would be substantially reduced and that further growth of work could be accommodated without a staff increase.

  The two computer manufacturers were invited to submit proposals. Company A suggested a small computer costing £16 000 for the hardware. They outlined the specifications of a tailor-made suite of programs which they would write and deliver with the machine within six months at a cost of £11 000. Company B suggested a machine costing £14 000. They recommended that the software should be obtained from service bureau SBX which had a ready-made suite of programs for supplier accounting and for control of stocks. This would cost £4000 and could be delivered in one month. The general manager contacted the service bureau, who confirmed the price and that this included one week's support from their staff to iron out teething troubles. They sent a substantial document describing the facilities, which seemed to cover the CABCO requirements although it was a highly detailed and tiring - not to say difficult - document to follow. SBX pointed out that if their standard system required amendments to suit this particular case, there would be additional charges at standard rates for the analysis and programming work involved. The general manager pressed them for a firm price on this, but they said they could not quote a firm price without doing a lot of the detailed analysis, for which they would have to charge whether or not an order was subsequently placed. This charge was estimated at £1700. However, they did say that their standard system was very comprehensive and flexible, and the cost of tailoring it to a particular case where this was needed was typically about £2000 and rarely more than £3000. The general manager was reluctant to commit £1700 when he did not know how the US parent would favour the whole deal.

  The general manager put Company B's proposal to the US parent company and submitted a report on the advantages to CABCO, mainly on the grounds of the clerical savings that would be made. What actually impressed the US management most was the improved timeliness and accuracy which was expected in making reports to the parent company on purchases from suppliers and on stocks in hand. These reports were vital to their world-wide strategy of supplier negotiation and materials stocking. They authorised taking up Company B's proposal. The general manager commissioned SBX to supply the programs and signed a contract for delivery of the computer.

  The machine was delivered in April, on schedule. At this time, a systems analyst from SBX came to explain the operation of the standard system. It worried the chief clerk that he could not understand how the computer would be able to produce the reports for the US parent company using only the data input to the standard system, but the systems analyst had a helpful, confident and experienced air and the clerk assumed that the analyst would be able to fix any deficiencies. It soon became apparent, however, that there were quite a number of deficiencies of the standard system in meeting CABCO's requirements, particularly as regards the special sequences and classifications needed to produce parent company reports. The method of inputting data did not match up with the existing manual procedures. Also, because of the unusually large number of suppliers, and other complications, an additional large effort was required to introduce a new parts coding scheme and supplier numbering system, and to capture these new details for the computer files. It took the analyst and the chief clerk about five weeks of hard work to agree all the changes that would have to be made and to specify the revised input and reporting programs.

Because the master files of information had to be changed in content and format, all the programs in the standard system required amendments. The service bureau estimated that the modifications would take two programmers two months to complete. They were told to get on with this as quickly as possible.

Because the chief clerk had been working practically full time with the systems analyst, there was a backlog of work in the existing system and a new temporary bookkeeper was taken on. The reports to the US parent, which were considered one of the most crucial responsibilities of the general manager, were delivered late and required frantic telephone conversations and explanations. The general manager was furious with this state of affairs and complained to Company B, who were demanding payment for their computer. Company B were apologetic and said that this was the first time they had had a customer have problems like this. The service bureau had been used successfully on more than twenty previous occasions. After a considerable argument, which was to involve the US parents of both CABCO and Company B, Company B eventually allowed a discount of 15% off the purchase price as compensation.

Meanwhile, towards the end of July, the programming work had encountered a few technical snags, but was expected to be ready by the end of August. It was agreed that a parallel-run test would be arranged at the end of September to prove the computer system by comparing its results with the manual system. Unfortunately, the chief clerk, who was not a young man, suffered a nervous breakdown as a result of the overwork and anxiety associated with meeting deadlines on the existing system, converting the supplier and part numbers and worrying about the new system, for which he felt he had considerable responsibility. He was the only member of CABCO who understood the details of both systems well enough to be able to specify what the criteria for the parallel run should be, and to know afterwards if it were correct. The general manager decided to delay the parallel run until October, when the chief clerk would have returned from sick leave and a holiday. The chief clerk returned, recovered, planned the parallel run with the help of the systems analyst and one of the programmers and, despite the increased workload (for which the bookkeepers were compensated by overtime payments), the two systems were run in parallel and the month-end results compared.

At this point things got worse. There seemed hardly a figure produced by the new system which matched the old one. The general manager complained to SBX and said that he was not going to pay further. They in return promised to get the system going according to the specifications without any further charge. They sent two people who took part in the review of the parallel-run results with the chief clerk and one of the bookkeepers. After about a week's work it had been demonstrated that differences in suppliers' accounts were nearly all accounted for by (a) a subtle difference in the cut-off rules which had been applied by the bookkeepers (these rules decided whether a particular transaction would be recorded in the present month's figures or deferred to next month) with the result that some transactions were recorded in one system but not the other, and (b) legitimate trans-actions which had been wrongly entered into the new system by the book-keepers and which had not been corrected, nor accounted for when making the comparisons. However, after all of the transactions concerned had been traced and brought into the calculation, they still did not completely account for the differences originally found. The reports for the parent company, some of which involved data which was not kept in accounting control, were wildly different.

Some of the control totals and proof listings produced by the new system were agreed as accurate. A makeshift arrangement was made whereby the data would be entered only into the new system, thereby accumulating it for future needs, but these intermediate totals and listings would be used by the bookkeepers to produce the accounts and reports manually. Meanwhile, production of accounts and reports was getting late, and parent company approval had to be obtained to advance the cut-off dates temporarily so that the prevailing deadlines could continue to be met.

The remaining errors in the suppliers' accounts were traced to a wrong assumption on the part of the systems analyst. This was corrected and the new system was allowed to produce suppliers' accounts from the end of December. Despite much midnight oil, though, the fault in parent company reports proved very elusive. A breakthrough came in January when a faulty assumption about the layout of data passed from a standard program to a special report program was discovered. This was a very obvious place to look for the fault and it had been closely checked by service bureau staff on more than one occasion, but they had all succumbed to the same false assumption! The result was that the long search for the fault was made in the wrong places. A by-product of this was that several examples of errors in the manual system's preparation of reports had been discovered. After the report fault was fixed, the new and old reports were much more similar, although they still differed in some details. SBX put this down to errors in the manual system. Eventually the general manager and the chief clerk came to accept this and to rely upon the new system, although nothing was ever proved (mainly because no-one was willing at this stage to take on the daunting task of the investigation that would be needed to decide one way or the other).

There were a number of other, less urgent, unsatisfactory features in the way the new system operated, but these were gradually fixed and by the end of June the new system was agreed to be completely operational. The temporary bookkeeper was released. The old cut-offs and deadlines were restored and slightly improved. The service bureau's bill for tailoring the system came to £7000 (additional to the £4000 purchase price of the standard system).

Asked if he did not think, in the light of this experience, that Company A's proposal had in fact been more realistic, the general manager replied, 'Yes, but had I put up their proposition, I don't think it would ever have been authorised, and we wouldn't have got our computer.' The tone of his voice clearly indicated that he thought failure to authorise would have been a mistake.

In August, the general manager received preliminary information from the US about new world-wide reporting requirements...but that's another story.

2   The comptroller of a plastics plant called in suggestions for mechanising his payroll system. He considered two proposals. One involved purchasing a minicomputer, the other involved sending the work to a service bureau. In the first case, there would be a purchase cost, including software, of

of about £14 000 and annual costs of about £2000, little affected by growth. Such an in-house computer would not be fully loaded and there would be potential for other work. In the second case, there would be annual costs of about £5000, increasing proportionally with growth in the number of employees, but no extra initial cost. The comptroller judged on this basis that purchase was the better bargain.

The plastics plant was a subsidiary of a large US multinational company, and the terms of local authority were that any capital equipment purchase over £5000 must be submitted to the parent for authorisation (even if it were actually financed by leasing). Expenditure on services and consumables was not subject to special authorisation, although local management was accountable for local profitability.

The comptroller decided 'to put the payroll processing onto the service bureau's computer.

ANSWER POINTERS

Section 2.1

1   It all depends how you define implementation, of course. Assuming you could not regard a system as 'implemented' while its accuracy was in doubt, then the diagram is properly drawn. I would define implementation as occurring when the new system is the one primarily relied upon to produce the desired results.

2

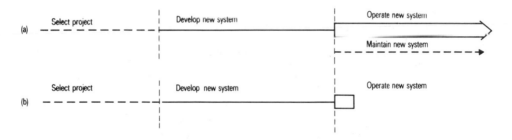

Fig. 2.8: Answer pointer to question 2, section 2.1

3   (a) Basically descriptive, but there are normative overtones since this is a tutorial book; I have explained what people do and I haven't implied that I disagree with this practice nor explained any alternative.

Another way of looking at the project control phases is to consider them a **prescriptive** model; a convention or ritual which people are expected to observe - maybe it helps to have a standard way of doing things, but it is not claimed that the particular way chosen is the best.
(b) Descriptive.

4   (a) There are a large number. Examples:
    style and cloth selection will be completely dependent on the tailor;
    the suit will not be sold on credit; there is no cash deposit required
    or this will be too small to need awaiting payday.
(b) See Figure 2.9 overleaf. Redundant dummy arrows in your answer are not important unless you are worried about elegance. Different dependences will probably mean that your assumptions are different from mine.

Section 2.2

1   The following list illustrates how these terms were used in the context

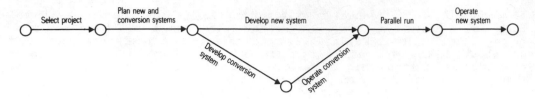

Fig. 2.9: Answer pointer to question 4(b), section 2.1

I found them. Some of the terms are so ambiguous or general that I can think of no good reason why your answer should be the same as mine. Items in brackets are names used for phases which are a smaller part of the ones I described – their position shows roughly where they were scheduled in their original context.

Project selection: project request.
Feasibility study: initial study, definition study, preliminary survey, system investigation, (current systems research).
Analysis: system proposal, requirements definition, functional specifications, (information analysis, outline design).
Design: system specification, program and procedure specification, detail design, system analysis and design.
Construction: programming, implementation, (program and human job development) (testing, proving) (installation, acceptance testing, system implementation).
Trial: system audit, post–audit, post–implementation evaluation.

  Possible reasons for difference:
a)   If larger projects need more phases, then additional names are needed for the extra phases.
b)   The phases are arbitrary and have little objective relevance to real-world activity, hence a lot of different interpretations are possible.
c)   Consultants and others offering advice on project control like to invent different terms in order to make a (phoney?) distinction between their advice and that of others.

2   If the consequences can be assimilated without a material increase in resources used, and if they do not materially worsen the estimated costs and benefits or the implementation time, then the new facts/requirements should forthwith be accounted for in the design.

  If the consequences cannot be assimilated and they vitally affect the success of the system, then the authorisation of the sponsor should be sought for the increased resources, reduced benefits, longer time etc., that will be entailed.

  If a new or changed requirement is not vital to the success of the system but cannot just be assimilated, the best strategy will usually be to continue to implement the frozen system unchanged and to schedule the requirement (assuming it is justifiable) for a system maintenance operation after the system is implemented.

3   Add a lead line from the tail of the 'test' rung to the tail of the 'analyse costs and benefits' rung. The lead line from the 'specify programs' rung to 'analyse costs and benefits' is superfluous and can be deleted.

Section 2.3

1   If having non–obvious advantages means that the sponsor will not authorise the project, it is a pointless project to propose. If the sponsor authorises a project with non–obvious advantages, but the lack of obvious advantages leads to lack of support from the users (or the sponsor) and

this in turn leads to loss of the desired benefit, or to later cancellation by the sponsor in the face of new or more obvious priorities, it was not a good project to have proposed.

2   If your answer referred to the post-implementation evaluation report, I disagree on two grounds. First, this report does not avoid the dispute; it can at best decide it. Secondly, if different assumptions or subjective judgements prevail, the report is unlikely to decide a dispute, if for no other reason than that the effectiveness of a system is not an objective fact.

  Avoidance of dispute will depend upon having agreement at the outset on unambiguous, objective measures of the target system performance. Objective, in this context, means that an observer will be able to observe or measure the performance when the system is operational, and that different observers will agree on the result. In practice, finding these objective measures may require some racking of brains, but the search is worthwhile. Examples of objective measures are as follows:

    to produce a report each month, within seven days of the end of the month, starting 1st January next, showing the ratio of goods returned value to goods sold value (the measure here is 'did the system do this, or didn't it?');
    to process all orders so that none is delayed more than three days before being received in the despatch department.

  The following type of measure can be called objective only if there will be enough information available to establish the facts:

    the cost per payslip is to be less than 20p per month;
    invoice preparation time is to be reduced to 24 hours on average;
    the cost of computer time used on the system is not to exceed £2000 per month;
    errors are to be reduced to less than 1 per 1000 invoices.

  It should be a target of a new system to provide the additional information needed to establish the measures of effectiveness. Objective measures embody targets expressed in terms of

    true or false (did it or didn't it?);
    time (to perform an operation);
    target date (for the system to be operational);
    cost;
    quantity;
    quality (where a quality target exists or quality can be objectively measured in true/false fashion).

3   Possible answer: finding out relevant facts; determining requirements; considering alternatives; setting objectives; designing computer procedures; designing business procedures; possibly developing and testing prototypes or pilot systems; predicting side-effects; estimating costs and benefits; planning; reporting.

4   Benefits – things that the sponsor desires.
    Costs – things that the sponsor wishes to avoid.
  If you accept these definitions, then you will have to use a phrase like 'financial costs and benefits' to describe money expenditures or income, and 'external costs and benefits' to describe things avoided or desired by persons other than the sponsor.

REFERENCES

(1) Benjamin, R. I., **Control of the information system development cycle,** Wiley-Interscience, New York, 1971.
(2) Gildersleeve, T. R., **Data processing project management,** Van Nostrand Reinhold, New York, 1974.

3 The most common computerised systems

3.1 ORDER PROCESSING

The purpose of these sections is to describe business procedures and requirements pertaining to those systems to which computers are most commonly
applied. The object is to illustrate the variety of choice which is possible
with even the most routine business systems, and to give insight into the
type of questions which must be asked if the facts relevant to designing
a new system are to be elicited. The first two sections are treated in much
greater detail than the subsequent ones. To maintain the same level of detail
throughout would make the descriptions as large as a book.
  Probably more than two-thirds of computers installed in business or public
administration organisations have as one of their applications, 'order
processing', i.e. identification of the goods or service required by the
customer or client, fulfilment of the order and charging for it. In a trading
organisation, 'order processing' is likely to be the recognised term for this
system - 'order entry', 'customer sales' are typical synonyms. Organisations
supplying services - banks, insurance companies, stockbrokers, government
departments such as social security - often have different names for these
procedures and a different terminology for the transactions and documents
involved. For example, a company trading in goods may send an invoice,
a broker may send a debit note, an insurance company may send a premium
renewal notice; but each fulfils a similar purpose. Although the names may
change, many of the concepts are the same or similar. From now on, the
terminology will be confined to that of a company supplying goods.

3.1.1 Identification of goods

The customer identifies the goods required by one of the following methods:

1 **Example**   The required goods, or an example of them, are indicated,
e.g. by picking up, pointing.
2 **Icon**   A photograph, drawing, picture, model etc. of the goods is
indicated or supplied.
3 **Natural language description**   A description in ordinary language is
indicated or supplied, e.g. 'a double-scoop plain ice'. Natural language
descriptions are usually rich in possible variety - different languages,
choice of synonymous words, alternative grammatical structures, idioms, etc.
4 **Formal description**   A stipulated or conventional description is indicated
or supplied. A formal description of the goods is an 'official' name used
to identify the goods. It may be one of the possible natural language
descriptions, e.g. 'vanilla flavour double cornet'.
5 **Code**   A string of symbols to identify the goods may be indicated or
supplied, e.g. part number '117461', model number 'A4A02'.

  Quite often, the trader displays the actual goods, an icon of the goods
or a natural language description of the goods, labelled with the formal
description or code. The customer need supply only the formal description
or code to identify the goods. The formal description is usually chosen to
be unambiguous and short, but it is not usually suitable for identification
of the goods in a computerised system. Computers work with symbols and

the symbol string of the formal description is quite likely to vary with alternative spellings, abbreviations, punctuations, upper and lower case, additional spaces. Codes, on the other hand, are usually defined so that no legitimate variation of the symbol string is possible (or, if variations are to be allowed, all possible variations are specified).

Often, the required goods have attributes which the customer or trader must define, or which are at variance with the chosen example or icon. These quality attributes may also be identified by natural language description, e.g. 'lightweight high-heeled' shoes; formal language description, e.g. 'court style' shoes; or code, e.g. style '12'. In addition, quality attributes may be described by a continuous variable. Although some qualities, such as shoe sizes or clothing sizes, are in fact continuous, in practice they are of course classified into standard sizes and so can be treated as **codes.** However, some continuous qualities are not so encoded and may be described to an agreed level of precision, e.g. the specific gravity of a consignment of oil may be '0.923'.

The **unit of measure** (continuous goods: e.g. litres, tonnes) or **pack size** (discrete goods: e.g. 24 x 1 kg cartons) must be declared or understood, together with the ordered **quantity** of such units or packs. When pack sizes are involved, customers are nearly always required to order an integral number of packs. Practice varies more widely with regard to goods sold by volume or weight.

3.1.2 Accepting the order

Figure 3.1 gives a flowchart of some of the processes typically involved in accepting an order. The 'Is it acceptable?' decision is merely to ask whether the order is understood and the company willing to trade in connection with the particular order. The two questions starting 'Do we' are intended to distinguish some companies' procedures from others' - but even a single company sometimes has different procedures for different types of goods or customer.

The flowchart is drawn for a type of sale that does not require pre-payment. Payment is then required at the time of supplying the goods or afterwards. Some organisations (e.g. mail-order houses) require prepayment for every transaction.

Where the goods are specially bought in or manufactured to meet an order, prepayment in full or of a deposit is quite frequently demanded. Prepayment for a back-order may also sometimes be expected.

Where a prepayment is expected, the terms of trade may be such as to require the customer to volunteer payment with his order, or it may be that an invoice is prepared and submitted to the customer, for payment before the order is accepted or the goods delivered.

The question 'Do we part-fill orders?' is also something of a simplification. Orders are often for more than one type of item, e.g. '100 widgets and 50 gadgets'. In this case, variations in policy for part-filling may be seen from the following examples. Suppose we have 50 widgets and 50 gadgets in stock. If orders are not part-filled at all, then (assuming we keep back-orders) we would keep the order until we get more widgets. (It may be that the existing stock of widgets and gadgets should be reserved for this customer.) If **orders** are part-filled, but **order items** are not part-filled, the 50 gadgets will be supplied and 100 widgets will be back-ordered (once more, reserving the existing widget stock should be considered). If order items are part-filled, the 50 gadgets and 50 widgets will be supplied and 50 widgets back-ordered.

Usually it is a matter of company policy whether or not back-ordering and part-filling will be done. One factor in deciding this policy is the effect on the complexity of the order processing system. Some companies who have decided to allow part-filling or back-ordering also allow their customers

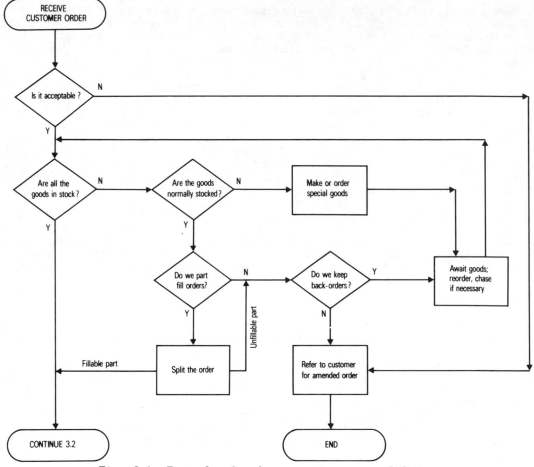

Fig. 3.1: Example of order acceptance possibilities

to elect, at the time of placing an order, not to have these facilities.

Question

1   Why, in a system which does not part–fill order items, might it be desirable to reserve stocks that part–fill a back–order?   (5 min)

3.1.2 Invoicing and despatch

The customer's credit is checked as in Figure 3.2. A credit limit is usually expressed in terms of the value of the present order, or the value of the present order and previous unpaid orders, or the number of days elapsed since the oldest unpaid order, or sometimes a more complicated rule combining age and value. Confirmation of acceptance of an order is usually sent if the customer may be in doubt about this, or if there is delay likely in delivering the goods.

At this point, there may be a pre–despatch invoicing system or a post–despatch invoicing system. The alternatives are illustrated in Figure 3.3.

In the case of customers who pay with their orders, preparation of an invoice is unnecessary, although in practice one may still be prepared.

When the goods are to be forwarded to the customer, the order must be made up by picking out the desired goods from the store or warehouse and

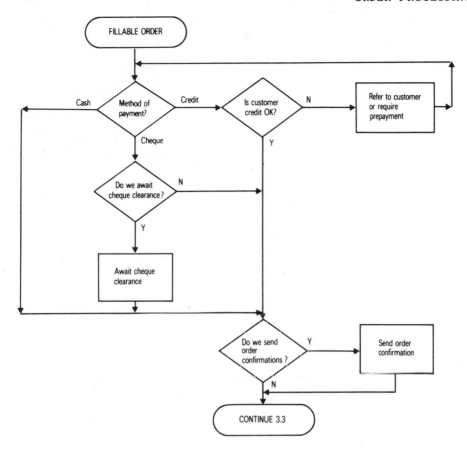

Fig. 3.2: Order confirmation

packing them up. A **picking note** is a document giving a warehouseman enough information to pick out the goods that make up an order. The picking note could be the original order, or a document specially written out, but it is most often a copy of the order (or invoice, in the case of pre-despatch invoicing systems). A **picking list** serves a similar purpose but has been prepared (usually by a computerised system) in such a way as to allow the warehouseman to fill several orders on one journey round the warehouse, thereby reducing the amount of travel involved.

A **despatch note** (delivery note, consignment note, goods outward advice, shipping note and other synonyms) is a document which contains enough information to enable the packed goods to be addressed and delivered to the consignee, who is usually, but not always, the customer. Usually a despatch note accompanies the goods and repeats the order details to allow the consignee to check that the consignment agrees with the order. Sometimes a separate shipping confirmation is sent by mail in addition. When the customer takes the goods, a despatch note is not needed. However, if a customer is to collect an order that was not filled from stock, he may have to be notified when it is ready.

Questions

2   What are the advantages and disadvantages of pre-despatch invoicing compared with post-despatch?   (10 min)

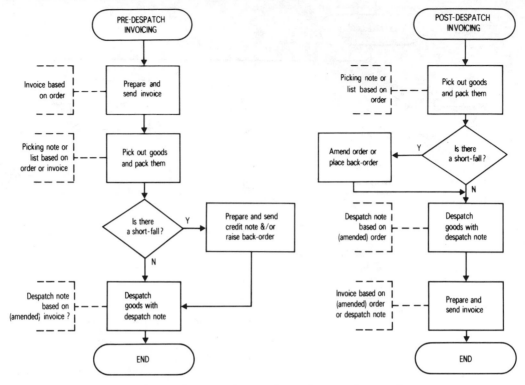

Fig. 3.3: Example pre–despatch and post–despatch systems

3   Sometimes a company wishes to send the invoice packed up with the goods. How would you have to amend (a) a pre–despatch invoicing system and (b) a post–despatch invoicing system?   (5 min)

3.1.3 The invoice

An invoice is illustrated in Figure 3.4. The fact that the delivery address and delivery instructions can be shown on this specimen suggests that it is a pre–despatch invoice, a copy of which is used as a despatch note. This is not conclusive evidence since it may be decided to put such inform-ation on a post–despatch invoice as a memorandum for the customer. Other possible copies (in a manual system) would be:

   a warehouse copy for picking and packing;
   an accounts copy for collection of the amount due and calculation of salesmen's commissions:
   a sales department copy for helping with customer enquiries and analysing sales.

   The quantity discount shown is an example of a class of discounts known as **trade** discounts. Other trade discounts may be allowed according to invoice value, particular products ordered, particular customer or type of customer. Usually, discounts are applied by following previously determined rules, but in some trades a discount rate is struck with each bargain. Trade discount is usually deducted when calculating the amount due shown on the invoice; an exception arises with a **retrospective** discount such as a discount allowed on a whole series of past invoices when in total they exceed some target quantity. Retrospective discounts can be handled by issuing a credit note.

INVOICE	**WIZARD WIDGETS LIMITED**	

New Works
Newtown

Customer	Shipping address (if different)	Our ref: 11760
⌐ A.Jones and Sons, High Street, Newtown.	⌐	Date: 1. 3. 80
		VAT Reg: 11746948

Your ref: — ⌐ ⌐

Special delivery instructions   Deliver to office, not factory

Units per pack	Number of packs ordered	Part number	Description	Number of packs shipped	Price per pack	Amount due
1	100	1174Y	Widgets, yellow	50*	1.00	50.00
1	50	0735W	Gadgets, white	50	2.00	100.00

Settlement discount : Deduct 2½% if paid within SEVEN days of the date shown on this invoice		
*Part-filled order: undelivered part will be shipped when stocks available	Less 5% quantity discount	150.00
		7.50 −
		142.50
	VAT @ 10%	14.25
	TOTAL DUE	156.75

Fig. 3.4: Example invoice

Settlement discount is a reward for early payment. It is earned only if payment is made within a designated time. If a customer pays in the allotted time, his account is credited with both his payment and the discount amount, so he will not be dunned for the difference. The discount amount (in the UK) comprises a Value-Added-Tax-portion (e.g. 15%) and a non-VAT-portion (e.g. 85%). The VAT-portion is recovered from VAT payments made to the tax authories.

Cash discount is a settlement discount accounted for when calculating the invoice total.

Export invoices usually bear additional information recording facts about supporting documentation, such as bill of lading, certificate of origin, marine insurance certificate, export licence, letter of credit. Overseas customers may also be treated differently in other respects, e.g. zero-rated for VAT.

A **credit note** is the opposite of an invoice and records a refund due to the customer, e.g. for retrospective discount, for goods returned. A **VAT list** is a statement for the tax authorities of the VAT collected on invoices and returned on credit notes.

Questions

4 If a settlement discount is allowed after a change in VAT rate, should

the new rate be applied in finding the VAT-portion of the discount, or the rate applied on the invoice?   (2 min)

5   If a retrospective quantity discount is allowed for the total quantities recorded on a series of invoices, some of which were issued before a change in VAT rate and some after, what rate of VAT should be used on the credit note?   (2 min)

6   A computerised system prepared invoices which invited the customer to deduct 2½% for early payment. It was found that there were sometimes penny differences in the discounts taken by the customer compared with the amount calculated by the computer. How could this be avoided?   (2 min)

3.2 CUSTOMER ACCOUNTING (SALES LEDGER)

The indebtedness of a customer to the company is recorded so that he can be dunned for payment. Dunning may take the form of repeat reminder invoices, often in increasingly strong language, or a statement summarising the invoices (and credit notes) issued and payments received (and made).
  Figure 3.5 illustrates a Balance Forward statement. The balance brought forward of £27.73 is the amount that appeared as 'balance due' on the previous statement. The new balance to carry forward is calculated from the invoices and credit notes issued, settlement discounts allowed, payments received and made and the balance brought forward.
  A Balance Forward statement is simple to maintain, but information about individual items is lost in the balance carried forward from one period to the next. If there is a large number of invoices and the customer pays according to his own calculations of the amount due, it can be a very complicated business requiring searching through old records if the customer does not agree with the balance claimed.

STATEMENT		**WIZARD WIDGETS LIMITED** New Works Newtown	

Customer

┌ A.Jones and Sons,
High Street,
Newtown.

                                          ┘   Month ended: 31st March, 1980

Invoice number	Date	Description	Amount
		Balance brought forward	27.73
11760	1. 3. 80	Goods	156.75
09442	12. 3. 80	Faulty goods allowance	10.00 −
11893	15. 3. 80	Goods	46.60
	15. 3. 80	Cheque - thank you	27.73 −
		BALANCE DUE :	193.35

Fig. 3.5: Example Balance Forward statement

An Open Item statement tries to ensure that this reconciliation is done constantly by carrying forward separately all unpaid invoices. Invoice items are deleted only when matched with a particular payment or payments. If a payment is received which does not match an outstanding invoice or invoices, the payment is recorded in the statement as an unreconciled item. It is then carried as a separate item until reconciliation is effected.

Sometimes statements are sent to an address other than the customer's address as it is stated on the invoice. Orders may be accepted from the separate branches of a large customer, invoices being sent to each branch, but statements are submitted to the customer's head office for payment.

When a payment is received, it is usually accompanied by a **remittance advice.** This may be a document prepared by the customer or a tear-off portion, or copy, of the invoice or statement. The payment is recorded in a cash book and paying-in slip for lodging with the bank, and the amount of the payment is recorded in the customer's account (with settlement discount where applicable). If Open Item statements are used, the payment is matched to the invoices it covers by reference to the remittance advice. If salesmen get commission and this is calculated on paid (as opposed to raised) invoices, commission can also be credited to the salesman at this point.

If a customer fails to pay, despite all attempts to collect from him, the balance of his account is written off eventually as a bad debt.

Questions

1   How could it happen that a payment is made to a customer who only **buys** from the company?   (2 min)

2   'Electronic Funds Transfer' is the name given to a system by which a large company can make payments instead of sending cheques. It is commonly used for making salary payments and for settlements of trade purchases.

The remitting company prepares a magnetic tape of 'transfers' showing the payee's bank and account numbers, payee's reference and amount payable. It sends this (in the UK) to the Bankers' Automated Clearing Service (BACS) which is a bankers' consortium. BACS makes the appropriate entries in the payer's and payee's accounts simultaneously on the day the payer designates. The payee's bank supplies him with details of the individual transfers made.

a)   Why does a company need the cooperation of the payees to use this service?

b)   How does the payer improve his own bank balance by using this system, although the payee's balances remain the same as they would be if payment were by cheque?

c)   Would it be best to make transfers relate to individual orders (invoices) or statement totals?   (5 min)

3   If salesmen are paid commission on raised (as opposed to paid) invoices, what question should the analyst ask when he is exploring the bad debt system?   (2 min)

3.3 MANAGEMENT REPORTS BASED ON CUSTOMER ACCOUNTING DATA

The following reports for management action are among those possible using customer accounting data.

1   **Overdue debts**   A list of debtors for 'chasing', together with information about their debts. Large lists may be ordered by age-of-debt or size-of-debt, or by size-within-age periods, so that priorities are clear. A summary statistic may be given by dividing total monthly sales into total customers' account balances, showing the average number of months' credit being taken.

SALES ANALYSIS: 12 MONTHS ENDED 31ST MARCH 1980
CONTRIBUTION–BY–VALUE BY SALESMAN
======================

RANK ORDER	SALESMAN NUMBER	NAME	SALES AMOUNT	CUMULATIVE	CUMULATIVE AS % TOTAL
1	1756	SMITH	127,252	127,252	25.4%
2	0973	ROBINSON	107,955	235,207	47.0%
3	0343	JONES	98,112	333,319	66.6%
4	0343	ADAMS	48,000	381,319	76.2%
5	5661	BAKER	22,119	403,438	80.6%
6	2773	FOX	20,200	423,638	84.7%
7	5212	CARPENTER	20,100	443,738	88.7%
8	0674	GRAY	19,900	463,638	92.7%
9	2111	HUGHES	15,101	478,739	95.7%
10	0788	DAWSON	10,212	488,951	97.9%
11	2222	JACKSON	10,100	499,051	100.0%

Fig. 3.6: Example contribution–by–value report

2   **Exceptional orders**   Orders over a certain value which may affect future production planning or sales strategy.

3   **Slow payers**   Identifying particular customers, or classes of customer, who consistently take longer credit than agreed. They can be negotiated with, or repeat business not sought.

4   **Quick payers**   Customers who take significantly less credit than agreed. They can be cultivated or rewarded with special terms.

5   **Sales analysis**   Categorising sales in any manner which is meaningful for management action: by customer or customer group; by salesman; by area; by product or product group; by product attribute. Reporting period requirements vary – this month, this month last year, last three months, previous three months, this three months last year, etc.

An **exception report** for sales analysis reports only those categories which have significantly changed (for better or worse) compared with a previous time period. The other categories are left unreported, since they do not need management action. (Some managers prefer to have explicit confirmation that things are changed little, even though such confirmation is not strictly necessary.)

A **contribution–by–value** report lists the sales categories in order of decreasing value and shows the proportion of the total sales that is contributed by the cumulative sales in the top categories (see the example in Figure 3.6, above). If 20% of salesmen account for 80% of sales, management may like to know which salesmen are in the 20%.

6   **Sales forecasts**   These seek to apply some rule to the past series of sales figures for a particular category, in order to predict sales over future time periods. This may be done simply by exponential smoothing (e.g. forecast for next month = 90% of the forecast made for the current month + 10% of the actual sales for the current month), in which case historic records need not be inspected. Alternatively, future sales may be predicted by fitting a curve to figures for past time periods and extrapolating this curve to the future. (A straight line model, the least squares line, is described in Appendix A.)

7   **Cash flow forecast**   The sales conribution to the company's bank balance in, say, the coming month can be estimated from the aged debts due and the average number of days credit taken in the past by customers. A more ambitious forecast would also take into account expected settlements on forecast sales during the coming period.

3.4 STOCK REPLENISHMENT

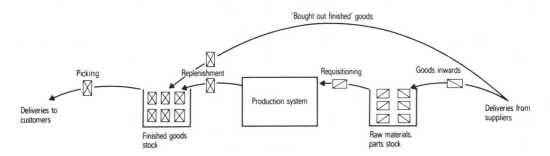

Fig. 3.7: **A simple model of production inventories**

Figure 3.7 illustrates two important inventories or stocks that are maintained in a production system where goods are manufactured 'for stock', as opposed to special order. Other possible inventories include a **spares** inventory, which holds a stock of parts for use when production machinery breaks down or wears out, and **work-in-progress** inventories which hold goods in an intermediate stage of production. Although the word 'inventory' strictly means a list which itemises stocks, the word is also popularly used as a synonym for stocks, as here.

Inventories may also be classified by geographical location. A distributor such as a supermarket chain may have only an inventory of 'finished goods', since all goods are bought finished, but this may be divided into inventories at individual supermarkets and inventories at regional warehouses from which local supermarkets are supplied.

As a stock of goods is diminished by picking or requisitioning, there is a risk of a 'stock-out', i.e. an inability to meet the orders because replenishments have not come in time. The effect of a stock-out may range from an isolated lost sale, or temporary rescheduling of work, to loss of a market or complete stoppage of work in a factory.

The stock level in a manual system is usually monitored on a **stock record card**, where the stock quantity in hand is marked down by the amount issued as a result of picking or requisitioning, or by the amount of stock scrapped because it is defective or obsolete (a **scrap note** is sometimes raised to record such scrapping). The quantity of stock remaining may be monitored at each issue to see if reordering is needed. This is sometimes called the 'two-bin system' because stock is reordered when the reorder level is reached, this state being indicated by the need to start on the second bin of stock. Alternatively, stock levels may be checked at regular time intervals: this method is sometimes called 'cyclical review'.

When stock falls below the reorder level, a replenishment supply is ordered in the hope that it will be delivered before stock falls to zero. The amount of the replenishment supply ordered at a time is called the **reorder quantity** and the time taken to replenish stock after a reorder is placed is called the **lead time**. Reorder levels and reorder quantities may be fixed arbitrarily in the light of experience. The well-known mathematical model for optimising these is outlined in Appendix A.

The recorded quantity of stock in hand may vary from that existing in the warehouse because of recording errors, unrecorded faulty stock, unrecorded exchanges for faulty stock supplied to customers, pilfering and theft, wastage, leakage etc. Sometimes these variances are surprisingly large. A **stock take** is usually undertaken periodically to count the stock physically and correct the recorded quantities.

Accounting to suppliers for bought raw materials, spares etc. is very similar to customer accounting. Copies of the order sent to the supplier perform a similar function to the copy invoices in customer accounting. Copy orders are usually held for matching with goods received (a Goods Inward Advice or Goods Received Note at goods reception is sometimes used for this matching). The supplier's invoice may need adjusting for shortfalls, defective goods etc. before the amount payable for the order goes forward to appear on the statement or remittance advice sent to suppliers.

3.5 PRODUCTION PLANNING AND CONTROL

These terms cover a variety of systems ranging from a comprehensive system for a manufacturing company, embracing sales and supplier accounting, control of inventories and long- and short-term planning, to a one-off simulation to find the optimum number of machines or operators for a given process. The operation of production planning and control systems is more dependent on the particular production process than the other systems described so far, so only a brief outline is possible here.

Production planning may refer to long-term planning to decide what products should be included in the product range and to estimate how much production capacity will be required. This may involve long-term forecasts of market demand and raw materials supply, as well as decisions on company policy and exploitation of technological alternatives.

In the medium term (say, a year), it may not be reasonable to contemplate changes of capacity or market, so these factors are considered to be constraints. Then the production planning problem is transformed into one of optimising the extent to which profit, sales or other output objectives are met in the medium term, subject to these constraints. The linear programming technique is widely advocated to aid this decision-taking (see Appendix A).

Production scheduling usually refers to short-term planning once a specific product mix or target inventory has been chosen; or, in a 'jobbing' type of factory where products are made to order, once customer orders have been placed. The problem is to **despatch** the jobs to the available production resources. This means planning which job is to be done next by each machine or each other resource in the factory (see Figure 3.8). Human despatchers faced with the problem of allocating known work among the men and machines available may adopt one of a wide variety of tactics. Finding a good despatching algorithm for a computer to follow can be a daunting proposition. For some types of job, a **network analysis** can be a useful tool. Computer **simulation** of the production process may also be a worthwhile exercise. An introduction to this topic, and a long list of references, can be found in **Computer Models in the Social Sciences**, by R. B. Coats and A. Parkin (Edward Arnold, 1977). Finding a good despatching algorithm is all the more important when shortage of production capacity makes efficient use of existing resources a priority.

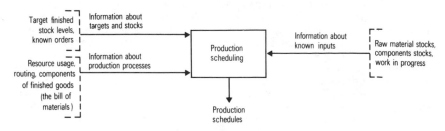

Fig. 3.8: Outline of a production scheduling system

Production control depends upon the collection of factory-floor information about the progress of scheduled jobs. This enables actual progress to be compared with the planned production, resulting in action to be taken to chase up the overdue jobs.

3.6 PAYROLL

Production of pay cheques and payslips for employees is usually considered in two parts: time-to-gross and gross-to-net.

The time-to-gross system is concerned with calculating the gross amount earned by each employee at the end of the week/fortnight/month. For hourly-paid employees, this will depend upon identification of basic and overtime hours worked and upon rate of pay. There are many possible variations resulting from shiftwork, piecework, bonuses, commissions etc. Holidays, sickness and absence may have to be accounted for, as will the receipt of insurance sickness benefits where these are deductible from pay. Overtime may also be payable to non-hourly-paid employees.

The gross-to-net calculation is more standard. In the UK, income tax is calculated by taking into account the employee's taxable pay in the tax year to date (after deduction of pension fund contributions), the tax paid in the tax year to date, the employee's income tax code and the prevailing tax rates. In addition to deducting income tax to find net pay, there may be deductions for national insurance contributions, superannuation or pension scheme contributions, and a wide range of voluntary deductions such as union dues, savings schemes, group medical insurance, social club fees. The payslip records these deductions together with the net pay due. Pay packets may be made up in cash, in which case a **coin analysis** will be required so that enough of each denomination of note and coin can be drawn from the bank to ensure that all pay packets can be made up with the exact change. Payment may also be made by cheque or by credit transfer direct to the employee's bank. Individual credit transfer slips may be deposited with single covering cheque at the employer's bank, for distribution through the normal banking system. In the case of large employers, Electronic Funds Transfer may be used (see section 3.2, question 2).

A payroll system also deals with incoming and leaving employees, changes in employee status or tax codes, changes in tax rates, payment of tax to the tax authorities, certificates of pay and tax deducted, as well as accounting analyses and management reports of payroll costs.

DISCUSSION CASE

Jokes about computers have had a good run. A particularly old and well-known joke, which brings a groan to the lips of most computer professionals, concerns the man who was being dunned by a computer for payment of an account of £0.00 and found that the only way he could stop the accounts coming was to send a cheque for £0.00. A few days later his bank manager rang him up to say, 'Please don't send cheques for £0.00 again; our computer can't handle them.'

Sometimes the facts are nearly equal to the fiction. I particularly like to imagine the red faces at The British Computer Society when it was discovered they had sent a subscription renewal invoice for £0.00 to one of their own members. In the United States, a well-known aircraft manufacturer had an automated supplier accounting system which also produced cheques in settlement of the accounts. Because of a program fault, a cheque for £30 000.00 was issued in error as £3 000 000. The supplier cheekily banked the cheque; one can imagine the consternation in the aircraft company.

Also in the United States, a book club had an automated system for invoicing members. The computer-produced invoices were mechanically burst, trimmed, placed in envelopes and franked for mailing. Unfortunately, a program fault caused each successive invoice to be printed with the same invoice details as the first. The unfortunate customer concerned had 100 000 invoices sent to him, each one in a separate envelope. This caused some difficulty in the small town in North Dakota where he lived, as the post office there had never handled so much mail before. However, they did eventually deliver the mail and the customer sent back one cheque for the $3 or so required, with a covering note to say he admitted defeat.

ASSIGNMENT

A group of entrepreneurs have decided to establish a mail-order wine club offering specially-imported wines to club members. Market research suggests that about 10 000 members could be expected in the first year, with perhaps five times as many in the longer run. Members would probably submit about four orders each per year, on average.

The idea is to have a small list of about fifty carefully-selected wines which will be stored in bins in a central warehouse. A member would identify on an order form how many bottles of which wine he requires. He would then calculate the amount of his bill, add carriage charges and send a cheque with his order. His order would be acknowledged and a fresh order form supplied. The wine would be picked from the bins to make up his order and the case(s) of wine despatched by common carrier or, for deliveries locally, the club's own van.

You have been asked to design a system which will process the orders and you have a chance now to ask the founders about their requirements for the system. Make a list of the questions you would ask. Details of the required system are given in the answer pointers; see if your questions would have drawn out all the facts listed there.

ANSWER POINTERS

Section 3.1

1   If existing stock is not reserved, it may be sold before replenishment stocks arrive. If the back-order exceeded the replenishment stock, it could not be filled and would have to await the next replenishment of stock – possibly ad infinitum! This could be annoying to a large customer, who is made to wait even though he can see smaller buyers getting their goods. A similar position can arise on the last back-orders if a series of back-orders can exceed the replenishment stock.

If orders are invariably small compared with replenishment quantities, the risk of this trap occurring may be considered too remote to warrant concern.

2   With pre-despatch invoicing: the picking/despatch note can be a copy of the invoice; earlier rendering of the invoice can lead to earlier payment for the goods; invoice preparation time is added to total delivery time, unless the picking/despatch note is based on the order or order confirmation, in which case invoice preparation and picking/despatch can proceed concurrently – but order confirmation time is added; it may be more confusing for a customer if a shortfall occurs; preparation of credit notes or amended invoices when shortfall occurs may add to delay; billing before delivery may go against trade custom; if delivery is interrupted for any reason, there is the embarassing risk of dunning a customer for payment for goods he has not been sent.

3   (a) Hold the invoice and marry the consignment with it, or use the invoice as a picking/despatch document as well.
(b) Hold the consignment and marry it with the invoice.

4   The rate applied on the invoice.

5   To be honest, the right **answer** to this question is not very important; I'm trying to illustrate the type of **question** that should arise in the mind of a systems analyst. Actually, in this case, the two rates applied on the invoices should both be used (according to VAT rules), the old rate being used for the discount amount attributable to the pre-change invoices, the new rate for that of the later invoices.

6   Show the discounted amount payable as a note on the invoice, not just the rate of settlement discount.

Section 3.2

1   The customer is invoiced for the goods. He pays for them. He later discovers them to be defective. The seller allows a refund by credit note. The customer has not made any further orders whose value exceeds this credit note. A payment is made to the customer to settle the credit balance in his account. Another possibility: overpayment.

2   (a) The payees need to supply their bank codes and account numbers and they also need to be willing to accept payment by credit transfer instead of by cheque.
(b) The payee's balance is normally increased only when a cheque is cleared. The cheque must have been paid in for presentation to the payee's bank a few days before the clearance. The payment will be debited to the payer's account some time before it is credited to the payee's account – currently about three days. For this period, the money is 'in the system' and not available to payer or payee.
   With Electronic Funds Transfer, the payer and payee entries are completed on the same day. The payer enjoys the use of his money for an extra three days or so.
(c) If a credit transfer matches an invoice, this makes reconciliation easier for the payee, but increases the number of transfers.

3   Should the commission that has been credited to the salesman be recovered on a bad debt? You may have worried over the continuation question: if commission is to be recovered, and statements are to be prepared on a balance forward basis, or statements are open item and contain an unreconciled payment at the time of write-off, how is the recoverable commission to be calculated, considering there is no identifiable 'unsettled invoice'?

Assignment
A quantity discount of $2\frac{1}{2}$% is to be allowed for five or more cases on one order. Carriage is to be fixed at £1.50 per case and 50p per case thereafter (part case counts as a whole case). Any differences between these charges and the charges actually made by the carrier will be met by the company. Obviously, the fixed carriage charges may be varied in the future. There are two carriers used: BNS and NVL. Customers may call and collect their orders if they prefer, in which case carriage is not charged, but this is likely in the case of only a small proportion of customers.
   VAT is chargeable on wine at 15%, although this rate may change in the future. The price per bottle shown on the order is to be inclusive of VAT. The system is to back-calculate from this the before-VAT price and the VAT payable. VAT is payable on carriage and is included in the above charges. The quantity discount is to be allowed on the wine only, not the carriage.
   The wine list will be revised every quarter or so. When a change in the list occurs, a revised order form will be circulated to all members. For

five days after posting the order forms, orders on the old list will be accepted, even if prices have changed, and will be charged at the old rates. (This is to accommodate members whose orders cross in the post with the new forms.)

If a member remits insufficient money to cover his order, his order will be returned with a request for the correct amount, except that if the short-fall is 30p or less, it will be written off (this 30p figure may be changed in the future). If the member remits too much, the excess will be kept in account and he will be invited to deduct this from his next order.

Orders for export will not be accepted.

The order is to carry a notice that if a desired wine is out-of-stock, the club will fill the order with the nearest substitute, at the price of the out-of-stock wine. Order items will not be part-filled. The club can choose the substitute and will normally choose the next highest price wine of the same colour and genre. The member is to be allowed to indicate that he will not take a substitute. It is not expected that wines will often be out-of-stock, since the policy will be to drop a wine from the list when stocks run low and sell off the bin-end through the trade. Although there will be about 20 suppliers of wine, they make their deliveries to two or three separately-managed bonded warehouses, from which the club draws replenishment supplies as required.

Picking is done into standard cases of one size only (twelve bottles). It is not expected that this size will ever change. Part-case orders are packed out with paper. The picking system has not been chosen, and the management is open to suggestions, but it would certainly be possible for the pickers to wheel a trolley carrying up to ten cases through the warehouse.

Each case is to be labelled with the destination. Only one delivery note is to accompany the cases, and it is to be packed into one case. The delivery note must state the number of cases in the consignment. A carrier's copy of the delivery note is to be handed to the carrier when he collects the consignment. The customer should be able to state special delivery instructions to be followed if there is no-one in (e.g. 'leave next door') and these instructions should appear on the carrier's copy. If goods can be reliably picked, packed and available for carriage within two days of the order acknowledgement, no confirmation of despatch will be necessary; otherwise a separate confirmation by mail will be needed. It is hoped that order acknowledgements will always be sent within ten working days of receipt of order, but they will be held for the first five days to await cheque clearance.

If goods fail to arrive or are damaged in transit, they will be replaced free. If there is no replacement stock and the customer declined substitutes, he will be credited with a refund of the original price prevailing (less quantity discount if allowed).

A fee of £2.00 is charged for a year's membership. This may change. Members may pay their renewal subscriptions with an order. Orders are not to be accepted from non-members or members whose membership has expired.

REFERENCE

(1)   Kennedy, J. N., **Errors in computer-prepared bills**, Industrial Engineering (USA), June 1977, pp. 30-3.

4 Requirements determination and fact-finding

4.1 SEARCHING FOR FACTS

What facts to find is a more important topic than the techniques used to find them. The analyst who understands what decisions he can or should ask for from the users, and what objective or quantified data he should accumulate, is in a strong position to make progress. Using his initiative he can likely get the desired facts even if he is ignorant of techniques. An analyst who is skilled in fact-finding techniques such as interviewing, taking samples or designing questionnaires is in a weak position if he does not know what facts he is after.

It is difficult, though, to make useful generalisations about what facts should be sought. Many books on systems analysis have prompting checklists for the analyst to review. These clues as to the type of question to ask are of little help in formulating a specific question. For example, it is difficult to see how a checklist question such as 'What operational alternatives are there?' can be translated into a set of more specific questions including, for example, for an analyst investigating an order entry system, 'Do you want a pre-despatch or a post-despatch invoicing system?' – unless the analyst already has insight into the operational alternatives. If the analyst has sufficient insight to formulate the more specific type of question, the value of the general prompt is at best only to trigger the specific questions. Such an experienced analyst would probably ask the specific questions without a prompt.

A rather similar position exists for the user who answers the questions. Faced with a general question like 'What operational alternatives are there?', where does he start to make an answer? Even if the user has insights into his system, maybe he does not understand the **possibilities** that are in prospect with a new system. A constructive discussion is far more likely to come if the user is asked a more specific question, like 'Do you want a pre-despatch or a post-despatch invoicing system?'

These points are to lend weight to the argument that checklist questions are not very helpful to the analyst who does not already have insights into the alternatives and possibilities of the particular system he is dealing with, and that they do not represent a useful starting point for his investigation. Maybe analysts are generally strong on conjecturing system/computer possibilities but are weak in knowledge of specific operational alternatives. Maybe users are strong on conjecturing operational alternatives but weak in knowledge of system possibilities. In such a case, **either** the ignorant analyst must discover the specific operational alternatives of the particular system he is dealing with, so that he can ask the user specific questions, **or** the ignorant user must be educated in computing and system design to the point where he can volunteer answers to general questions from a strong position of knowing the possibilities. From the sponsor's point of view, concerned with the overall effectiveness of the investigation, neither of these options should be neglected. In giving terms of reference, the sponsor should make clear what responsibilities/training he envisages for analyst and user. I shall continue this explanation on the basis that the analyst has the responsibility of discovering operational alternatives. In practice, many experienced analysts assume this responsibility if the sponsor is not specific and the users appear uninformed.

If the analyst is going to discover specific operational alternatives, his main strategies are:

1   external search for equivalent systems;
2   model underlying processes of the specific system;
3   local search in and around the existing system.

These are explained in the following sections and are subjectively listed above in order of likely decreasing ratio of discoveries made to effort expended. All three strategies may be followed in a given case. An analyst who has previously investigated other systems similar to the one now under consideration will scarcely want strategy 1. An analyst who has detailed knowledge of the physical processes involved in the sections of the business concerned may ignore strategy 2.

In addition, there are some general conceptual alternatives in providing information for decision-taking by managers. I believe it is valuable for an analyst to understand these alternatives before exploring user requirements. They are described in section 4.5.

Some of the facts the analyst collects are **private**, in the sense that they help him to do his work but are not of especial interest to others, perhaps because the others already know those facts. The personalities of the participants is an example. Other facts are **public**, and must be shared with or communicated to others. Private facts may be stored only in the analyst's head; public facts need to be documented.

Questions

1   The phrase 'the systems analyst determines the users' requirements' is nicely ambiguous as regards the word 'determines'. What are the alternative meanings, and which meaning describes the preferred behaviour of the analyst?   (5 min)

2   A fact-finding checklist I have before me has a list of 90 general questions, followed by:
    91 What else must be looked for?
    92 What has been missed?
 Of what possible value are these questions?   (5 min)

3   The examples of this section emphasise the type of facts which might be classified as **user requirement** facts, i.e. those which record the operational alternatives or preferences of the sponsor and user managers, what output is required and what business procedures are to be followed. Other facts which the analyst seeks could be classed as **evaluation** facts for the sponsor and **design decision support** facts for the analyst. Speculate on the definition of these and give two examples of each.   (5 min)

4.2 SEARCH FOR EQUIVALENT SYSTEMS

There are many thousands of computers in operation throughout the world, in many different types of industry, doing many types of job. The chances of a proposed system not having a close parallel which has already been investigated are quite slim. A day or two spent finding and studying the details or history of a close parallel system is likely to repay itself many-fold in that it may help to

a)   provide prompts for the analyst's fact-finding questions,
b)   suggest alternatives, and
c)   avoid pitfalls.

The sources to consider are as follows:

1   Computer manufacturers' representative   Most large manufacturers employ specialists who may be able to advise directly on a proposed system or to direct the enquirer to an example. The qualifications of the adviser should be weighed up before accepting advice.

2   Computer users' groups   These are self-help consumer associations formed by purchasers of computers. There is an association for each major computer marque. Exchange of information between members is usually encouraged (even competing companies often cooperate).

3   Packages   A program package already designed for an application can be a good source of ideas, even if the package is not suitable. If the package facilities match requirements, there may be savings in development costs. Information about packages can be obtained from **The CUYB Directory of Software** (published annually by Computer User's Year Book), from the Software List service of the National Computing Centre and from bibliographies published by computer manufacturers and user groups.

4   Trade or industry association, government departments   An association designed to promote or research matters of common interest exists for most trades. Government ministries and the Local Authorities Management Services Advisory Committee (LAMSAC) fulfil a similar role in public administration.

5   Literature   There are useful descriptions to be found in published books, reports and articles, but tracking them down may call for some skill or experience. **Trade literature** (package descriptions, manuals, brochures, reports written by trade associations, etc.) can be an especially rich source, but (apart from government reports) is usually poorly stocked by libraries. It is often best traced through the sources listed above. General literature such as is usually stocked by technical libraries can be traced through the library catalogues, through bibliographies (which are usually kept in the reference area), through abstracting journals purchased by the library (e.g. Anbar Data Processing Abstracts, ACM Computing Reviews, Computer and Control Abstracts, Data Processing Digest, NCC Computing Journal Abstracts) or by using a computer-search service which will turn up the references in a computer-held catalogue which satisfy the search criteria (references which contain stated keywords in the title or abstract). A number of **books** contain case studies, generally of the more common types of system. A short guide to books containing fairly substantial case studies is given in Figure 4.1 overleaf.

Question

1   Your local authority is considering keeping a computer database of all land parcels and properties in its area, in order to provide information about the consequences of alternative highway and other development plans. They wish to consider digitising the spatial data about the properties so that a computer search of properties affected by a plan can be made, and they wish to consider presenting the information in graphic form. What publications in your library are relevant, what packages are relevant and where have similar projects been undertaken? (60 min after reaching library)

4.3 MODELLING THE OBJECT SYSTEM

Assuming the analyst is to take an active role, it is essential that he has a good personal model of the processes of the business, or the portion of it under investigation, in terms that are meaningful to him in a real and tangible sense. If he lacks this model of how the mission of the organisation is pursued in terms of real objects, physical locations, movements and events, he is in no position to evaluate what he is told about the information needed to guide the organisation's activities. If he cannot evaluate

INDUSTRY/APPLICATION	A	B	C	D	E	F	G	H	I	J
Bakery	5		5				5	5		
Ball bearing manufacturer							6	6		
Banking	6									1
Clothing, wholesale					1					
Commercial laundry products								7		
Communications								6		
Computer sales										5
Construction, civil engineers										6,7
Copper					2					
Domestic utensils					5					
Drugs, wholesale							4			
Electromechanical equipment	5									
Electronics					1		3			
Equipment hire							5			
Food, wholesale			1				1,6	2	1	
Footwear, retail									5	
Gardening equipment								5		
Glass						6				
Greetings cards							1		1	
Groceries, retail							1		1	
Hardware, wholesale			7				7		7	
Hire purchase				6						
Industrial gases	6							6		
Investment										7
Leisure	2									
Machine tools								5		
Newsagents, books, retail	6	6		2			6		6	
Power tool manufacturer								1		
Power utility				1						
Public houses	6									
Publishing	2			2	2					
Textile yarn making								2		
Timber products			1							
Trade union	1					1				
Unspecified								7		

KEY

1   Lucas, H. C., and Gibson, C. F., A casebook for management information systems, McGraw-Hill, 1976.
2   Higgins, J. C., Information systems for planning and control: concepts and cases, Edward Arnold, 1976.
3   Science Research Associates, Case study in business systems design, SRA 1970.
4   Science Research Associates, Case study II: Medco Inc., SRA, 1973.
5   Clifton, H. D., Data processing systems design, Business Books, 1971.
6   Lambourne, S., Computer applications in business, Longman, 1974.
7   Lucas, H. C., Information systems concepts for management, McGraw-Hill, 1978.

A	Accounting/financial information	F	Membership/personnel records
B	Accounts payable	G	Order processing
C	Accounts receivable	H	Production and strategic planning
D	Distribution	I	Stock control
E	Marketing information	J	Miscellaneous: Tender estimating, negotiable instruments, portfolio management, mutual funds.

Fig. 4.1: A short guide to data processing case studies

what he is told by managers and other users, he is practically committed to playing a passive role. His model is a private fact and need not be documented.

Modelling this object system has two other values. First, it may show up loose ends in the analyst's understanding of the processes and dictate further facts to be found. Secondly, it may help the analyst identify the important decisions or control actions so that he can give independent consideration to the information needed to support such decisions or actions (this is sometimes called 'decision analysis').

For illustration, consider an order processing system. The **object** system has to do with goods and customers; storing goods in bins, customers choosing goods, men picking and packing goods into containers, men and vehicles delivering goods to customers, customers sending payments in exchange for goods, storing payments, deliveries of replenishment goods from suppliers. Items such as a customer order form, an invoice, a delivery note, are part of the **message** system which describes, regulates, coordinates or controls the object system. These messages are not desired in themselves, but are called for by present methods of fulfilling the underlying mission. (You may be objecting that 'payments' above may be messages in the form of bank notes or cheques or other tokens such as coins, and they are not 'object' things. This is a valid point, and if you insist on taking it into account then you must describe the object system in terms of the ultimate exchange of goods which the banking system supports by the messages embodied in cheques and notes. Assuming you already have a personal model of how the banking system substitutes for physical exchanges, however, and since it is a subsystem common to many you will investigate, it is a convenient fiction to regard cheques, notes, etc. as things in the object system desired for their own sakes.)

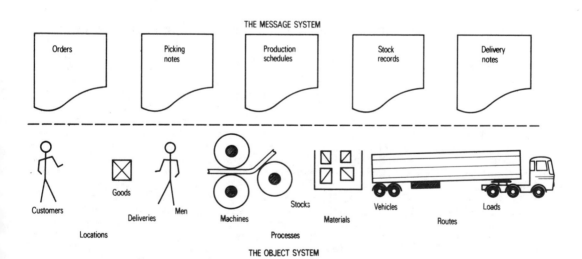

Fig. 4.2: Message system/object system

The object system may be modelled in a number of ways – a map, drawing, block diagrams, precedence networks. The important thing is that the analyst arrives at a model which is personally explanatory to him. This may be a purely narrative or mental model.

A convenient starting point is to describe the tangible outputs of the object system (the goods or services which it is the organisation's mission to supply), the inputs used to produce them and the transformations (production and distribution processes) by which the outputs are derived from the inputs. It is often a problem to decide how much detail to include. This problem can be rationalised by identifying a dozen or so orthogonal components of the system at first, then using judgement to decide which of these to divide into sub-components, and so on, from the top down.

The object system outputs and inputs referred to above do not generally include data output and input. In some systems, where it is the organisation's mission to provide financial or information services, it may be convenient to regard those messages as belonging to the object system. For example, in an insurance company the policies and claim payments may be considered part of the object system, whereas supporting messages such as premium notices and statistics are part of the message system.

Questions

1   Define 'object system'. (5 min)

2   Describe broadly the object system of the mail-order wine club (assignment at the end of chapter 3). What gaps are there in your knowledge of how the object system is to work? (10 min)

3   Is it practical to design a message system purely on the basis of a model of the object system and a top-down analysis of the decisions and actions involved in it? (2 min)

4   Are the terms 'message system' and 'data processing system' synonymous? (2 min)

5   The personnel department of a police force administers a message system, the police personnel records system. What are the main components of the object system? (2 min)

4.4 LOCAL SEARCH IN THE EXISTING MESSAGE SYSTEM: BOTTOM-UP ANALYSIS

Although conceiving or observing the object system can be a cost-effective means of fact-finding and can lead to original ideas for data processing systems, unfettered by existing practice, examination of the existing message system is usually considered a prerequisite of designing a replacement. This is because the existing messages, and the procedures associated with them, are likely to reveal exceptional cases and requirements which may be overlooked in a top-down analysis. Investigation of the existing system may be omitted when a participant of the systems analysis team is judged to have sufficient all-round knowledge of its operation to be able to describe it without investigation, or where existing procedures are already adequately documented.

This fact-finding is conceptually accomplished by identifying all the formal messages which leave the system or which are regarded as an end-product in themselves (e.g. information for managers, data stored to help answer future queries) and tracing back how these messages are derived from messages which enter the system or from data captured within it. Whether the systems analyst starts by looking for outputs or inputs is in practice immaterial as long as he ends up with a complete description.

So, the analyst's target is generally to find all the documents and other formal messages (such as messages on visual display units) presently in use in the system and to describe the procedures by which the output messages are derived from the input messages. The analyst must ask the following questions.

1   What is the purpose of this data, information or report? (Maybe it has none, it served an old need and is not required in a new system.)

2   What new information or reports are desirable to improve the system? (Maybe these can be prepared using data already present in the system.)

The analyst may also consider what new information or reports could be produced from the data of the system, and submit his ideas for review by the users. A dummy or mock-up report is likely to be more meaningful to users than just a description.

Methods of finding out about the existing system and user requirements by interviewing, observing, etc. are described in chapter 5. The facts and requirements turned up will form the basis of subsequent proposals and may be used by many people other than the analyst. A standard method of recording these public facts is covered in chapter 6. Analysing the data used in the system, and planning the organisation of it to meet the require- ments of the proposed system, are covered in chapter 7.

Throughout this chapter, the emphasis has been on finding impersonal facts about the system. Through contact with users in his search for such facts, the systems analyst will also get equally important impressions of user preferences, abilities and other personal characteristics which may be important to the eventual design. Some of these aspects are developed in chapter 8.

4.5 ATTRIBUTES OF INFORMATION

Information characteristics can be measured on a number of dimensions. Some are intuitively obvious to most people, for example 'summation', the accumulation of a total of transactions over some time period, or 'aggreg- ation', the lumping of information into categories, e.g. sales by product, sales by salesman. Some less obvious characteristics are:

   age of information ('currency'), dictated by report interval or frequency
      and report delay;
   sample size and dispersion;
   distortion, particularly by money values and seasonal variation;
   accuracy or error;
   precision (of numeric information).

Age of information   Condition information describes the state of affairs as at a particular instant of time. A balance sheet reports account balances 'as at 30th June 1980'; a stock report gives stocks in hand as at the end of the week. The age of condition information is always at least equal to the delay between making the observations and reporting the facts. This is called the reporting delay, **d**.

Condition information periodically reported to management refers only to specific points in time, such as the end of the week or the end of the year. The manager does not have condition information about intermediate points in time. The time between successive points to which a report relates is called the report interval, **i**, and is the reciprocal of the report frequency. A report produced with an interval of one-twelfth of a year is produced at a frequency of twelve times a year. Figure 4.3 illustrates the concepts.

Consider a manager who wishes to know the current state of affairs at time **B**. He will just catch the report for week 4 so he gets the newest inform- ation, which is only **d** old. Consider now his seeking information at time **A**. He can only have the previous report, that for week 3, so he faces the oldest information, which is **d+i** old. If his requests for information were to be uniformly distributed in time, then the **average** age of the condition information in the reports he inspects would be the mid-point of the oldest and newest ages, i.e. $d+\frac{1}{2}i$.

Operating information is a summation of a number of transactions over a period of time. A profit and loss account reports income and expenditure

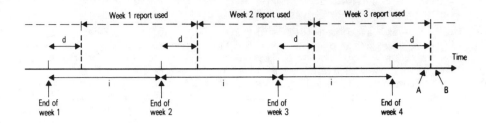

Fig. 4.3: Reports of condition information (see page 45)

for 'the twelve months ended 31st December 1980'; a sales analysis report totals sales during the previous week. The reporting delay, **d**, refers to the delay between the **end** of the period to which a report relates and the time of production of the report. The report interval, **i**, is the interval between the end periods of a regularly-produced report. The period of time over which the report sums the transactions is the report period, **p**. The report period is often the same as the report interval, but not always. If **p** is greater than **i** (for example, profit and loss over the preceding twelve month period reported at monthly intervals), the report is called a **rolling** report, since conceptually the oldest month is being rolled out of the sum and the newest month rolled in. (If **p** were less than **i**, the reports would be summing a sample of the total transactions.) Figure 4.4 illustrates the concepts.

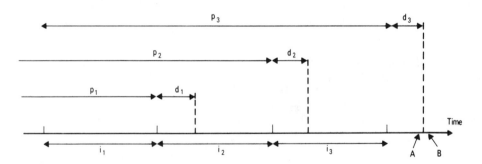

Fig. 4.4: Age of operating information

When the manager seeks information in the best case, point **B**, he gets information at time d_3 after the end period of p_3. However, the transactions summed in the report occurred throughout the period p_3. If we consider the age of operating information to be equal to the average age of the transactions making it up, then assuming the transactions occur uniformly in time, the minimum age of operating information is $d+\frac{1}{2}p$.

In the worst case, when information is sought at point of time **A**, the previous report, covering period p_2, is the most recent available. The end point of this report's period is $d_3 + i_3$ old and it also sums transactions which are on average $\frac{1}{2}p$ old; so the maximum age of operating information is $d+i+\frac{1}{2}p$.

Again, if the manager's needs for information arise uniformly in time, the average age of his operating information would be at the mid-point of the oldest and newest: $d+\frac{1}{2}i+\frac{1}{2}p$.

Sample size and dispersion   When A. Jones and Sons placed their order for 100 widgets, were they stockpiling against an expected widget shortage, or have they over estimated their need for widgets? In either case, they may not re-order for a long time. Is the economy picking up, or have Jones found a new use for widgets? In either case, they may re-order more quickly than usual.

Without knowledge of Jones's motives, it is impossible to infer a trend from a single order. Statistical inference is the art of making inferences when knowledge of specific causes and reasons is lacking, or is too complex to be accounted for. The number of observations – the sample size – is important to the credibility of the inference. If you toss a coin twice, and it comes down heads on both occasions, you should not infer that it is double-headed. If you toss it 100 times, with heads the result each time, maybe you should. Likewise you cannot infer much from one or two orders for 100 widgets in a month, but with 100 such orders you may feel justified in inferring that widget demand is not likely to be nil next month, assuming it is not panic buying in advance of expected scarcity.

Dispersion is the name given to the variability of a number of observations about their mean. If, for example, daily sales are **widely dispersed**, you would find that a number of separate day's takings spanned a wide range of values and were not closely grouped around an average day's takings. If daily sales are widely dispersed, you can infer virtually nothing about trends just because today's data happens to be widely above or below average. If monthly sales, for example, are **narrowly dispersed** and the present month's figures are widely above or below average, there is more reason to suppose that they reflect a trend or some out-of-the-ordinary event which could call for management action.

When a manager uses an operating information report, or a series of condition information reports, as a basis for action, he is making a statistical inference. There is some evidence that managers who lack statistical training do not have a good intuitive understanding of the impact of sample size and dispersion on the reliability of the inference. This gives an argument for reporting trends (moving averages) and variances on previous periods; see left side of Figure 4.4 overleaf.

Distortion   Distortion refers to the way information, even though accurately reported in the sense that no mistake is made, may mislead the user if he fails to apply some correction to it. Distortion is a predictable bias of the true facts which, it is supposed, are the ones sought by the user of the information. The term 'bias' is typically used for distortion created by humans; for example, biased estimates may be supplied.

An insidious and gross form of distortion found throughout financial management reporting is the distortion of money values. Briefly, even though the quantity of goods and services produced in an economy may remain the same, an increase in the money supply will increase the prices people as a whole are willing to pay for those goods and services. This is inflation. If we consider the tonnage or volume of goods produced and the numbers of people receiving the services as being the 'true' facts of interest, we must declare that a report of money values paid is a distorted picture of reality. People who would not dream of adding numbers of kilograms to numbers of pounds weight will happily add money value of sales this quarter to money value of sales last quarter, in a time of inflation.

Two approaches to correcting this distortion can be adopted. One is to report information in unchanged units of weight or volume, etc. A possible difficulty here is that changes of **quality** may bring distortion into these figures. Number of cars made this year compared with number of cars made last year is a distorted comparison if this year's cars are an improvement on last year's. The second approach is to 'deflate' the reported money values to their 'real' values by scaling them by a measure of general prices (a price index). A possible objection here is that there is no reliable index, but government published indices may be considered a reasonable approximation. (There will never be a completely reliable index, reality

		a	b	c	d	e	f	g	h
			Last	Trend	Vari-	Proportion	Variation	Exp.	Adjusted
		Sales	12	(ave.)	ation	of trend	trend	var.	sales
Year	Month	(tons)	months	=b÷12	=a−c	=d÷c	=.9f_{12}+.1e	=fxc	=a−g
1978	Jan	42	3368	281	−239	−0.851	−0.851	−239	281
	Feb	27	3365	280	−253	−0.904	−0.904	−253	280
	Mar	33	3368	281	−248	−0.883	−0.883	−248	281
	Apr	101	3373	281	−180	−0.641	−0.641	−180	281
	May	220	3388	282	−62	−0.220	−0.220	−62	282
	Jun	563	3446	287	276	0.962	0.962	276	287
	Jul	649	3475	290	359	1.238	1.238	359	290
	Aug	1027	3512	293	734	2.505	2.505	734	293
	Sep	473	3485	290	183	0.631	0.631	183	290
	Oct	202	3457	288	−86	−0.299	−0.299	−86	288
	Nov	125	3482	290	−165	−0.569	−0.569	−165	290
	Dec	19	3481	290	−271	−0.934	−0.934	−271	290
1979	Jan	8	3447	287	−279	−0.972	−0.863	−248	256
	Feb	29	3449	287	−258	−0.899	−0.904	−259	288
	Mar	40	3456	288	−248	−0.861	−0.881	−254	294
	Apr	201	3556	296	−95	−0.321	−0.609	−180	381
	May	623	3959	330	293	0.888	0.109	−36	659
	Jun	722	4118	343	379	1.105	0.976	335	387
	Jul	989	4458	372	617	1.659	1.280	476	513
	Aug	1246	4677	390	856	2.195	2.474	965	281
	Sep	404	4608	384	20	0.052	0.573	220	184
	Oct	99	4505	375	−276	−0.736	−0.343	−129	228
	Nov	37	4417	368	−331	−0.899	−0.602	−222	259
	Dec	10	4408	367	−357	−0.973	−0.938	−344	354
1980	Jan	50	4450	371	−321	−0.865	−0.863	−320	370
	Feb	77	4498	375	−298	−0.795	−0.893	−335	412
	Mar	118	4576	381	−263	−0.690	−0.862	−328	446
	Apr	398	4773	398	0	0.000	−0.548	−218	616
	May	399	4549	379	20	0.053	0.093	−35	434
	Jun	1017	4844	404	613	1.517	1.030	416	601

Fig. 4.5: Report of monthly sales. Detailed figures are shown to reveal the working. The seasonally adjusted sales, h, are derived by subtracting the amount of 'expected variation', g, from the actual sales, a. When h is above the trend, c, this shows a better-than-expected month; when h is less than c the month was worse than normally expected. This report uses an exponentially-smoothed variation trend which for the sake of the example has been initiated in 1978; obviously with only three years' experience to go on, the latest figures cannot yet be considered reliable.

is too elusive.)

Another common distortion is that of seasonal variation. Although associated particularly with seasonal businesses such as ice-cream selling, seasonal repercussions are felt in nearly all businesses. Indeed, whole economies have regular seasonal fluctuations. The fluctuations that affect particular businesses may be caused by the weather, by feasts such as Christmas, or by other less obvious factors such as the end of the official tax year. A reasonable seasonal correction can be made by removing the expected amount of fluctuation from the actual figures. This is done by recording the variations from the trend in previous periods, expressed as a proportion of the trend in the period, and using this proportion over a number of years to make a forecast of the amount of variation in the figures for the current period. After a number of 'seasons' have been experienced, so that the variation trend may be considered reliable, a substantial difference between the seasonally-adjusted figures and those of the trend reflects a significant event.

Accuracy   Information is accurate if it is free from error caused by mistake, data incorrectly recorded at source, error in transcription or processing and unrepresentative sampling. It may be known that errors occur, but if their sense or magnitude is unpredictable then no correction can be applied in the way it could with distortion. The most that can be done is to report the likely range of error (the limits of accuracy) if this can be established, or to give some other indication of confidence in the reported figure. Note that figures may be internally accurate in the sense of being correctly accounted for, but wrong in representing reality. For example, the total of sales this month may correctly reflect the total of the invoices issued this month, but it is inaccurate (does not reflect reality) if one of the invoices issued contained an incorrectly calculated total or wrong number of goods delivered. It is only accuracy in relation to reality that should be important to the manager. Establishing internal accuracy is just a step on the way to this real accuracy.

Precision   Precision refers to the number of significant digits supplied in reporting a figure. Establishing internal accuracy may require maintaining records precise to the penny, but managers may be satisfied with much less precisely reported information, perhaps to the nearest thousand pounds. There is not much point in reporting management information with a precision finer than the limits of accuracy. Indeed, doing so can give a misleading impression of accuracy.

Questions

1   If a manager complains that he needs more up-to-date information from periodic reports, what can be done to help him? What would be the effect on the value of the information (increased/reduced/no change) for making inferences?   (10 min)

2   The salesmen receive precise reports of their weekly sales in money values, even though the dispersion and distortion in these reports is so large as to preclude any inference from the latest figure. Yet the sales manager still wants them to get the reports, and the salesmen still want to receive them. Why?   (2 min)

3   Accountants may be excused for adding this month's money value of sales to last month's, in a time of inflation, but not managers. Why?   (2 min)

4   (a) A Retail Price Index, published monthly, is established with January 1969 = 100. The current value is 237.4. Let the value of the index be x_i where i is the month 1, 2, 3...n (n being the current month).

  Deflating a report to 1969 values produces figures which are related to an arbitrary 'real' point. These are meaningless in terms of today's values. You have been asked to design a report which gives today's values for earlier months. You have each x_i available. Assuming v_i is the money value of a figure relating to month i, write an algebraic expression which will give you R_{ni} , the present value of any given month's figure, in terms of v_i and x_i .

(b) In Figure 4.5, what are the seasonally-adjusted sales for July if actual sales in July are 1357 tons?   (15 min)

5   A safety limit, in a system which is being controlled, is a value which a variable in the system should not reach. The safety line is the value which should not unknowingly be reached, if it is desired that the safety limit is never passed. The difference between the safety line and the safety limit, called the safety margin, should be the maximum disturbance that can occur on the variable (in the direction of the safety limit) during the sum of the report interval, report delay and any time involved in counter-

acting the disturbance by taking remedial action. Thus, in a cyclic review stock control system, where stocks are reviewed every 28 days, where the report is given to the stock controller 2 days after the review, where it takes 30 days to re-order, and take delivery of, replenishment stocks and where the safety limit is zero stock, the safety margin is the maximum total stock issues that can occur in a 60 day period.

(a) In a stock control system, how can the safety stock level (the safety line) be reduced?   (5 min)

(b) A car driver checks the distance to the car in front every second. At speeds below 10 metres/second, both cars can be braked to a standstill virtually instantaneously. It takes 0.1 seconds, after sighting the distance to the car in front, for the driver's nervous system to interpret this image and decide whether or not action is required. It takes a further 0.6 seconds to apply the brakes if necessary. Supposing all this data were true, how far behind the car in front should the driver keep, when travelling at 7 metres/second, if he wishes to avoid collision?   (2 min)

DISCUSSION CASE

A subsidiary company was required by its parent to send a monthly state-ment of its trading. The manager of the subsidiary found that the parent company's queries about the increase or decrease in sales etc. were a nuisance. He felt he was on top of things and didn't like having to explain the monthly variations. With his colleagues, he devised a ruse to diminish the queries. This entailed 'smoothing out' the fluctuations in the reports that would otherwise be sent to the parent. If sales were high, for example, the excess quantity would be deducted from the reported figure and credited to a suspense account carried to the next month. If sales were low, sales in the suspense account were released to the report, or, if the suspense account was empty, a crash programme would be organised towards the end of the month to make sure every outstanding transaction was processed. The managers made sure that the suspense account was empty at the end of the financial year, so the annual accounts were perfectly correct.

ANSWER POINTERS

Section 4.1

1   Meaning 1: The analyst chooses/decides/dictates/spells out the require-
      ments.
      Meaning 2: The analyst discovers/finds out/is told the requirements.
 Of course, the final authority to decide the requirements rests with the sponsor or his delegate, so assuming authority has not been delegated to the analyst meaning 2 is the only one to hold water as regards **authority.**
 On the other hand, meaning 2 describes a **passive** role for the analyst which is unlikely to be effective unless an **active** role is being taken by the sponsor or other user. If for any reason the sponsor/user does not take an active role, e.g. lack of motivation, desire or knowledge, it may behove the systems analyst, on his own initiative, to promote knowledge of altern-atives and, if no one else is willing, to promote a definition of specific requirements which he believes best fit the criteria. This behaviour is closer to meaning 1.
 Throughout the rest of this book, requirements are considered as if through the eyes of an analyst who is 'finding out'. In the companion book, Systems Management, requirements are considered as if through the eyes of a manager who is 'deciding'.

2   Little, unless they are effective as general 'set-breaking' exhortations.

Psychological or perceptual **set** is the name given to the fixed approach we tend to have when problem-solving: a sort of predisposition to solve a problem in a particular way. When deeply involved in a problem, we tend to fasten on to alternatives close to the things we already have under consideration; more rarely do we break out onto entirely fresh ground by a flash of insight or 'lateral thinking'. The well-known nine-dot problem:

. . .

. . .

. . .

(join all the dots with four straight lines without lifting pen from paper) is first approached by most people by exploring lines originating and terminating on the dots. Many people get **set** or fixed on this general method of solution and cannot solve the problem, because the only solutions require considering lines which extend beyond the figure. Psychological set is exploited in many puzzles and riddles. Try this old one: 'Two Americans were crossing a bridge. One was the father of the other one's son. What relation were they?' (Answer at end of chapter.)

If the analyst was 'set' in a narrow approach to fact-finding, a large checklist may help break this set and encourage him to consider new possibilities. On the other hand a checklist itself represents a particular view of the problem and method of solution. It encourages a new 'set' where the analyst explores only the possibilities suggested by the checklist. If the analyst takes questions 90 and 91 seriously, as exhortations to re-consider his approach, there is some chance (slight?) that they may be responsible for a new idea.

3 Evaluation facts: ones which will help evaluation of broad alternatives, certainly by the sponsor and maybe by others. They provide information which supports the feasibility decision and selection from major alternatives. Example: the difference in financial costs of operating the existing and proposed systems; the size of a backlog of work observed in the present system; operational trends; the estimated time to develop a new system; and so on.

Design decision support facts: ones which will help evaluation of finer, technical alternatives. They provide information which supports computer database and program design decisions. Examples: the distribution of the volume of orders per day; the number of customers; the distribution of the number of characters in customer names; growth trends of volumes of processing; and so on.

During the feasibility study stage, the analyst is searching for all the **evaluation** facts, most (if not all) of the **user requirement** facts and the most important **design decision support** facts. At the end of the analysis phase, he hopes to have all the user requirement facts and most (if not all) of the design decision support facts.

The evaluation facts which are required are considered in detail in Systems Management. Which design decision support facts are required comes out of knowledge of, or speculation about, alternative design possibilities, which are treated later in this book. For now I shall continue to emphasise user requirement facts, while readily conceding that this division is somewhat artificial because a single fact may be classifiable under more than one heading.

Section 4.2

1 Perraton and Baxter (Eds), Models, evaluation and information systems for planners, **LUBFS Conf. Proc. 1**, MTP Construction Press, 1974, especially pp. 231-8. (Source: abstracts.)
Broadhurst, C. J. C., Spatial retrieval for point-referenced data, **Department of Environment Research Report 18**. (Source: catalogue.)
SYMAP, National Computing Centre. (Source: CUYB Directory of Software.)

POINTMAP, Applied Research of Cambridge. (Source: CUYB Directory of Software.)

Newcastle upon Tyne County Borough; Woodhouse, Smith and Reay, 1971. (Source: Department of Environment Register of Research 1974, found via catalogue.)

Cleveland County Council, Teare, 1976. (Source: Research and Surveys 1976, published by HQ Library, Department of Environment, found via catalogue.)

Section 4.3

1   The object system is the system referred to by the message system. The message system communicates messages about the object system. (Another author calls the object system the 'referent' system. This is an apt name, but it hasn't caught on.)

2   Obviously your answer is unique to you. My crude sketch:

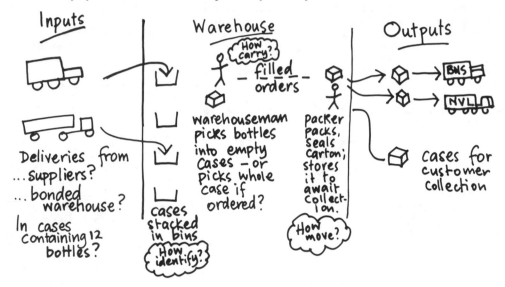

Fig. 4.6: Mail-order wine club object system

Some of my gaps/questions which turned up as a result of thinking about the object system: How is it decided which carrier is going to deliver the goods? Can the same case in which the wine arrived from the supplier be used for delivery to customers? What methods will be used for carrying the cases while picking, moving picked cases to packing and moving packed cases to await despatch? How will warehousemen know which bins to use for storing new supplies delivered? Is there any chance of mechanical equipment to automate picking and packing, or is that too fanciful?

3   Possible, yes. Practical, maybe. Desirable, probably not.

The mail-order wine club doesn't yet exist and if the formalised part of the message system (the bit the analyst is asked to design) is to exist before it starts business, there is little alternative but to design the system on the basis of a mental model and top-down analysis. The risk in this procedure is that something important will be overlooked. This risk has serious consequences in computerised systems, which cannot usually be changed quickly. Where there is already an existing system, the risk can be reduced by investigating the messages actually used in the existing system.

4   I think not. In most common use, 'data processing system' refers to a **formal** system using messages of predefined types, i.e. messages comprising strings of symbols whose general format has been defined. Naturally, there are many other messages passed in an organisation, in the form of ordinary human discourse, and these comprise an informal, undefined system. 'Message system' includes both formal and informal messages.

The formal system of passing messages in a manual system corresponds to the idea of 'recognised paperwork'. When a recognised document is prepared (e.g. order form, invoice, delivery note) I like to think of this as the preparation of a **formal message**. What should go in the message has been predefined. This book is all about such formal messages, and when I use the term 'data processing (DP) system' I mean a system for preparing, sending, receiving these formal messages.

5   The policemen, their deployment, their training, their promotions, and so on.

Section 4.5

1   Reduce report delay – no change in value, apart from the increased value due to reduced age.
    Reduce report interval (increase report frequency) – no change in value.
    Reduce report period – reduced value resulting from smaller sample.

2   Maybe the sales manager sees the report as a motivator, maintaining salesmen's morale when sales are up and spurring them to greater effort when sales are down. Maybe the salesmen want to check their commissions. Maybe everybody is just curious.

3   The accountant may deal with money values to check the internal accuracy of the data processing system. Assuming the manager is satisfied about internal accuracy, he should concentrate on the information's relation to external reality.

4   (a) For this month, $R_{ni} = v_i . (237.4/x_i)$ or, in general $R_{ni} = v_i . (x_n/x_i)$.
(b) Last twelve months' sales up by 1357–989 = 368 tons to 5212 tons. Revised trend is 5212÷12 = 434, so variation from trend is 1357–434 = 923, or 923÷434 = 2.127 as a proportion of trend. The previous variation trend for July was estimated at 1.280, so the latest forecast is 0.9x1.280+0.1x2.127 = 1.365. The expected variation amount is therefore 1.365x434 = 592, and the seasonally-adjusted sales are 1357–592 = **765**.

The exponential smoothing used has given undue weight to the 1978 figures. This could be improved by using a different forecast, e.g. mean variation proportion to date or perhaps smoothing with less weight given to the past, as a temporary expedient until more years' experience is to hand.

5   (a) By lowering the safety limit. By reducing the report interval – possibly to zero, by using a two-bin system where stock level is checked every time it is changed. By reducing the report delay. By reducing the replenishment time.
(b) 11.9 metres.

Answer to riddle p. 51: Husband and wife. Just in case you didn't get set, I'd better explain that most people get set on the idea that the two were males – and no doubt they are preconditioned to expect that this sort of question has a complicated answer.

REFERENCES

(1) Butterworth, J.E., **The accounting system as an information function**, Journal of Accounting Research, **10**, 1, 1–27, 1972.

(2) Staff of EDP Analyzer, **What information do managers need?**, EDP Analyzer, 17, 6, June 1979.
(3) Stamper, R. K., **Information in business and administrative systems**, Batsford, London, 1973.
(4) Taggart, W. M., and Tharp, M. O., **A survey of information requirements analysis techniques**, Computing Surveys, 9, 4, 273–90, December 1977.

5 Fact-finding techniques

5.1 PLANNED DISCUSSIONS

These are usually called 'interviews' by systems analysts, but if a phrase like 'the analyst interviews the user' is taken in the same sense as 'the reporter interviews the politician' or, worse, 'the employer interviews the job applicant', then this word interview is suspect. An analyst who interviews in this sense is assuming the active role and may, through his approach, draw the user inintentionally into playing the passive role.
  This is not to say that there is never a place for an interview led or dominated by the analyst, but the analyst ought to think twice before conducting an interview in which he keeps such tight control over the reins. It is through points of analyst–user personal contact during private discussion that the user has most practical opportunities to influence the system design. Although committees, staff meetings, newsletters and presentations are manifestly participatory, they rarely engender the same sense of involvement as comes from a face–to–face discussion between analyst and user. The sponsor is usually a powerful person as regards decision–taking and formal authorisation. The analyst is in a strong position to influence his decisions, which effectively gives the analyst a share in the sponsor's power over the users. Users are often powerful as regards the success or otherwise of the operation of that part of the system with which they are concerned. If the analyst gains their confidence then he can share in the user's power to influence the sponsor. Potentially, the analyst is a focal point for reconciling the conflicting interests of different groups.
  Each person whom the analyst interviews is likely to have an interest in the design or operation of the system. Maybe the analyst should ask himself, of every such person, 'Am I planning to interview someone who should be participating in a more active way? If so, what should be the extent of his participation?'

5.1.1 Planning discussions/interviews

The best order for a series of discussions and interviews is probably from top management down, but this is not vital. In practice, the difficulty of tying down some busy managers may oblige the analyst to accept any sequence, time and place he can get. It is usual protocol to make appointments with management and technicians before having interviews with them, and to clear the lines with departmental management before approaching shop floor or clerical workers. If the interviewee is given an indication of the expected duration of the interview, e.g. 'not more than an hour', this gives him a chance to plan his day around it and to arrange freedom from interruption.
  Many people are sensitive about the authority and status they enjoy. The analyst should pause to reflect before contacting a manager 'out of the blue' to fix up an interview. A happier outcome is likely to result if the first contact is made by the manager's manager – a strategy that will be aided by a top–down interview plan. It can also be helpful if the manager's manager is present at the initial meeting, or at least attends long enough

to provide an introduction, to make responsibilities clear, and to give some endorsement to the importance of the exercise.

Before the discussion, the analyst should prepare an 'interview guide'. This is a prompting checklist or agenda which the analyst can use to ensure that he raises all the points he intended, assuming these points are not pre-empted by earlier answers and that there is enough time to raise all the points. If an interview exceeds the allotted time, it is sensible to adjourn and reconvene at a freshly appointed time. This avoids disrupting the day and allows both parties to reflect on what are obviously not short, straightforward matters.

5.1.2 Conducting discussions/interviews

Experience suggests that just telling an interviewer what mistakes are possible in interviewing does not stop them happening. The inexperienced interviewer, perhaps, has too many things on his mind to prevent mistakes. The experienced interviewer, whilst probably less error-prone, sometimes remains oblivious to the few faults he has. It is a worthwhile exercise for the trainee analyst to interview a role-playing interviewee, who can use the checklist in Figure 5.1, or his common sense, to advise the interviewer on his performance.

Did the interviewer:

make a prior appointment, state expected duration;
try to get a sensible time and place;
introduce himself agreeably;
state his purpose in a way you understood;
state what he believed your role to be;
appear to have planned the discussion, state the agenda;
ask questions appropriate to your knowledge and status;
listen to your answers and opinions;
quiz you on uncertain terminology or vague answers;
question obvious things as well as the less obvious ones;
ask for sets of documents and completed specimens;
make notes;
check back his facts or understanding before leaving;
depart agreeably, leaving the door open for a return visit?

Did he:

arrive late;
stick unduly to his plan;
annoy you through mannerisms, excessive jocularity, etc.;
interrupt your answer and prevent you from completing your point;
allow you too much rein with opinion or side-tracks;
confuse you with technicalities;
argue with you or criticise your suggestions;
dominate when he should have given you more freedom;
anticipate your answers or jump to conclusions;
part with missing detail or understanding?

Fig. 5.1: Interviewee checklist on interviewer performance

5.1.3 After the interview

It is easy for interviewee or analyst to get a wrong impression from an interview. Error can be reduced if the analyst sends the interviewee a resume of the matters agreed or facts recorded; see the example in Figure 5.2 opposite. This confirmation can also clarify responsibilities for further action – failing this, there is greater risk that agreed action will not get

Discussion Record NCC	Title BAKERY ORDER ENTRY		System HOE	Document 2 1	Name REQU13	Sheet 1

Participants BK, SW, MBL	Date FEB 1 1980
	Location HO
Objective/Agenda DAILY SCHEDULES	Duration 45 MIN

Results:	Cross-reference

1. Production schedules must be ready for the night shift by 5.30 p.m.

2. Amendments to standing orders for next day's deliveries are on hand by 4.30 p.m. at the latest.

3. Acceptability to van drivers of optical mark read document, for taking new standing orders and temporarily or permanently amending old ones, to be investigated.

<u>Action by</u>   BK

4. Example of OMR document in use in a similar application to be obtained and circulated.

<u>Action by</u>   SW

S21

Author
SW

Fig. 5.2: Discussion record

done. The analyst should in any case write up his notes immediately after the interview.

The analyst should reconsider his interview plan and judge whether or not to accept the picture given by the interviewee or whether to get confirmation from another source. It can be a mistake to accept one person's description of what another person does with data. It can also be a mistake to accept a person's description of organisation policy reasons for following a procedure when that person does not formulate policy.

Question

1   The following are events I have often observed in the course of interviews by systems analysts. Speculatively, what are the possible causes of each event and what should the analyst do in each case?

a)   The interviewee, possibly suspicious of the analyst's motives, does not cooperate or wish to participate.

b)   The interviewee seizes the opportunity to put forward his pet solution or opinion.

c)   The analyst does not fully understand the interviewee's explanation of a complicated document or procedure in the system, even though he asks the interviewee to repeat it once or twice. Eventually, he feels he cannot ask for any more repetition and leaves, still not understanding.

d)   The analyst interjects a fresh question before the interviewee has finished his answer to the last one. The interviewee never gets back to the topic to deliver an important piece of information. A similar event occurs when the analyst fails to ask a good follow-up question.

(30 min)

5.2 DOCUMENT COLLECTION, SAMPLING, OBSERVING

Documents carry useful information about the data elements and terminology used in the existing system. The analyst should aim to collect a completed specimen (or photocopy) of each relevant document, together with its copies, and to record relevant facts such as who completes the document, how it is routed, how it is filed. Figure 5.3 shows a specimen of the National Computing Centre's standard form for recording those facts, although it should not be thought that this form exhausts the possible facts of interest. Preparation of a document/department routing grid analysis (Figure 5.4) may reveal gaps in the analyst's knowledge of document use. A document/ data element grid analysis (Figure 5.5) can reveal the need for information about how data elements on output messages are derived from data elements input, or about the use to which non-output data elements are put.

Sampling is quite often used to quantify questions of average (What is the mean time to serve a customer from receipt of order to despatch of goods? What is the average number of amendments to standing orders per day?) or of proportion (What proportion of customer accounts have more than ten transactions in a month?). Usually, the 'observations' for the sample can be obtained from an existing body of records (filed copy orders, filed and dated despatch notes, filed copy statements and so on) but sometimes the analyst may have to make his own observations or introduce a variation to a procedure to generate the data he requires. Usually, the decisions which the sampling may affect are not very sensitive to inaccuracy, so in the interests of expediency a rough-and-ready, rule-of-thumb approach to sample size and sample selection is often adopted. Appendix B contains some formulae that may prove useful when a little more rigour is required.

Observing is not usually aimed at answering specific questions in quite the same way as in sampling; it is more commonly a browsing activity.

Clerical Document Specification N C C	Document description WIZARD WIDGETS INVOICE (see Fig 3.4)		System WOE	Document 4.1	Name INVOICE	Sheet 1
	Stationery ref. —	Size A5		Number of parts 4	Method of preparation Typed	
	Filing sequence Acc. Copy — Inv. No. Sales copy — Cust/Inv.No	Medium NCR Set		Prepared/maintained by Invoice dept.		
	Frequency of preparation/update As required	Retention period Accounts - 3 yrs Sales - 1 year		Location Accounts Dept. Sales Dept.		

MONTHLY VOLUME	Minimum 250	Maximum 600	Av/Abs 350	Growth rate/fluctuations Peak March/April Growing 15% p.a.

Users/recipients	Purpose	Frequency of use
#1. Customer 2. Accounts 3. Sales 4. Picking/ Despatch	1. To bill amount due 2. To update customer ledgers 3. To answer customer queries 4. To pick out and route goods	1. Each order 2. " " 3. Once per 1000 4. Each order

Ref.	Item	Picture	Occurrence	Value range	Source of data
1	Customer name	X (40)	1		Order
2	Customer address line	X (40)	0-3		Order or cust. record card
3	Shipping address	X (40)	0-4		Order
4	Invoice number	9 (5)	1		Invoice # register
5	Invoice date	99.99.99	1	English date DD MM YY	
6	Customer's reference	X (40)	0-1		Order
7	Delivery instruction line	X (60)	0-2		Order
8	Pack size	999		1-144	Parts catalog.
9	Packs ordered	9 (5)		1-25,000	Order
10	Part number	9(4)A		See parts cat.	order or parts catalog.
11	Description	X (17)	1-4		"
12	Packs shipped	9 (5)		1-25,000	Order, possibly amended
13	Price per pack	9(4).99			order or parts catalogue
14	Amount due	9(5).99			calc
15	Total before discount	9(5).99	1		calc
16	Quantity discount %	99.99			
17	Quantity disc. amt.	9(4).99-	0-1		calc
18	Total after discount	9(5).99			calc
19	VAT rate %	99.99	1		
20	VAT amount	9(5).99	1		calc
21	Total due	9(5).99	1		calc

Notes

[The Ref. numbers are marked alongside the items on a specimen attached]

S 41

Author M	Issue Date	6.3.80

Fig. 5.3: Clerical Document Specification

Title				System	Document	Name
WIZARD WIDGETS DOCUMENT / DEPARTMENT GRID				WOE	5.1	DOCDEPT

Departments \ Document	Customer purchase order	Sales order	Invoice (#1)	Invoice (#2)	Invoice (#3)	Invoice (#4)	Consignment note (#1)	Consignment note (#2)
Sales manager	1							
Order office	2	¼	2	2	2	2		2
Stock control	2							
Typing	3	1	1	1	1			
Warehouse					3			
Despatch					4	1	1	1
Accounts			3					
Post		3						

Fig. 5.4: Document/department routing grid

Browsing through existing records maintained in user departments aids the analyst's understanding and can also reveal facts which may not become apparent by other means. The most productive observation comes when the analyst has a guide who can draw out the features of the system. As well as supporting the verbal descriptions he is given, observation may reveal things which may not arise during discussion; backlogs, extremes of noise in working conditions, physical layout and pace of movement, clutter and so on. These facts may help the analyst when he comes to consider the effectiveness of alternatives by imagining how a proposed system would work in the physical environment.

Question

1   You are planning a new order entry system which is intended to speed up order processing and despatch by cutting down the existing average time by two days. Orders presently take a widely varying time to fill. The slowest cases take about 40 days, ten times longer than the fastest. There is no time recorded for order fulfilment in the present system, but the new system will record the time taken to process each order and report the average. Approximately how many representative orders should you time through the present system so that you can quantify the eventual improve-

Chart Sheet NCC	Title: WIZARD WIDGETS LTD DOCUMENT / ITEM GRID	System WOE	Document 5.1	Name SALESDOCS	Sheet 1

Documents / Items grid

Items \ Documents	Customer Purchase order	Sales order	Invoice	Consignment note	Parts catalogue	Stock record	Customer record	Statement	Invoice number register
Customer name	I	T	O	O			M	O	
Customer address	I	T	O	O			M	O	
Shipping address	I	T	O	O					
Invoice number		T	O	O	M		M	O	M
Invoice date			O	O			M	O	
Customer's ref.	I	T	O	O					
Delivery instructions	I	T	O	O					
Pack size	I	T	O	O	M	M			
Packs ordered	I	T	O	O		M			
Part number	I	T	O	O	M	M			
Description	I	T	O	O	M	M			
Packs shipped		T	O	O		M			
Price per pack		T	O		M				
Amount due		T	O						
Total before discount		T	O						
Quantity disc amount		T	O						
Total after discount		T	O						
VAT amount		T	O						
Total due		T	O				M	O	
Statement total								O	
Statement date								O	

I = item on the document when it enters the system
O = item on the document when it leaves the system
M = item on a master record which remains within the system
T = a transfer item, entered on a document which is not a master record and does not leave the system

S34

Author ℳ	Issue		Date 7 3.80

Fig. 5.5: Document/data element grid

ment made by the new system? (Refer to Appendix B.)   (20 min)

5.3 QUESTIONNAIRES

Questionnaires are an unsatisfactory way of collecting facts or opinions for business system design. The analyst will generally do better to make personal visits, in which case the questions potentially in his questionnaire go into his interview guide instead. Questionnaires have the following failings:

> low response may mean unrepresentative sample;
>
> reliability (will the same person give the same answers on two different occasions?) is often untested or untestable, but face-value errors are common;
>
> validity (does the question elicit the intended information?) is often untested or untestable, but face-value errors are common;
>
> the limitations of written questioning may lead to a shallow question being asked, or to a shallow answer being supplied, with no follow-up to ensure that the response is accurately interpreted;
>
> people may be prejudiced against questionnaires (they receive too many, they mistrust the method, they are sceptical of the use that will be made of them, they believe them to be depersonalising) and as a result ignore them or complete them rashly, misleadingly, or untruthfully.

Judging from my own experience of completing questionnaires, few designers of questionnaires understand the principles of questionnaire design. Natural ability to design questionnaires is either lacking or very thinly spread in the population.

As a last resort, a questionnaire may be the only practical contact when large numbers of far-flung people are involved. The following steps, prior to distributing a questionnaire, are recommended for questionnaire designers.

1 Read Berdie and Anderson (1).
2 State your precise objectives in sending the questionnaire. What decisions will the questionnaire influence? What different decisions will be taken for different hypothetical questionnaire results?
3 Estimate the time and cost of conducting the study.

If still interested:

4 Plan a questionnaire with a personal tone, which has a purpose which will be clear to the recipients. Plan how you are going to process the responses.
5 Ask short questions which call for short, objective answers (but make sure you are not asking for a shallow conclusion on a topic that deserves an essay).
6 Test the questions by showing them to your peers.
7 Test the revised questions by asking a small sample of the target population to complete the questionnaire blind (i.e. with no more explanation or personal contact than is planned for the other recipients). Repeat the questions in interview to elicit difficulties, ambiguities, unreliability.

Question

1 What is or could be wrong with these questionnaire questions?
a) (To a local government officer)
   Why did you think of entering local government?
      (i) Advice from school   (ii) Advertisement (iii) Parent's advice
      (iv) Friend's advice   (v) Youth employment advice

    (vi) Other (please specify) ....
b) (To a branch manager)
   What suggestions do you have for improving the effectiveness of branch
   order processing systems?
c) (To a branch manager)
   What is annual branch expenditure on stationery consumables?
d) (To a sales representative)
   Would you prefer a pre-despatch system for your customers?   (10 min)

ASSIGNMENT

Read the Hartree Bakeries case in Clifton, op. cit. Figure 4.1.

ANSWER POINTERS

Section 5.1

1   (a) Possible causes: the interviewee's awareness of computers from the
media/other departments/own experience is negative, and he is fearful of
what they bring for him; the interviewee is antagonistic to the management
or to the organisation; the interviewee has not been briefed adequately prior
to the interview; the interviewee has good grounds for his suspicion; the
interviewee prefers not to cooperate or participate, perhaps because he
cannot spare the time, perhaps because he does not realise the possible
consequences of failure to participate, perhaps because his alternative
activity is more enjoyable or important to him.
  Some of the causes suggest, in turn, that the sponsor has failed to arrange
the right climate for change in the organisation, or has misjudged the
climate, or has misjudged the need for cooperation and participation. The
analyst is unlikely to be able to reverse deep-seated suspicions in a short
time-scale. If the whole-hearted cooperation of the interviewee is important
to the operational success of the system, then maybe the analyst's best
course is to recommend a temporary stop to progress while the sponsor
establishes the right climate. This may take courage!
  If the interviewee is just uninformed, not hostile, maybe the analyst can
correct this (but it should have been done beforehand). If the interviewee
does not appreciate the importance of his participation (assuming it is
important, which it usually is) maybe the analyst or sponsor have not
spelled out possible effects of the new system on organisation achievement
and personal job satisfaction.
(b) Possible causes: the interviewee is an opportunist who wishes to
ingratiate himself with someone whom he believes influences management;
the interviewee's idea or opinion is a good one and corrects an operational
weakness he has detected; the interviewee's idea or opinion is not a good
one but is well meant; the interviewee wishes to prejudice the analyst
against rival ideas or opinions which are against the interviewee's
interests.
  The analyst does well to listen to ideas and sound out opinion. He need
not shut them out for fear they will corrupt him. Misconceived ideas or
opinions will need reconciliation in the eventual system, just as will better-
found ones. The analyst will eventually hear rival ideas or opinions, so
unless he is good at eating his own words his safest strategy is not to
endorse publicly those he is offered in the early stages.
(c) Possible causes: the analyst does not understand the terminology of the
interviewee; the analyst lacks a conceptual model of the object system onto
which he can project the explanation; the analyst is missing some other
aspect of the message system which is important to understanding this

aspect; there is too much detail to assimilate in one go. Play dumb. Make a joke of it if overstretching. Go away and return later.
(d) Possible causes: the analyst is too intent on following his interview guide and is not listening to or evaluating the interviewee's answers; the analyst is dealing only in words, i.e. he is not mapping the words onto a model of the object system, possibly because he lacks the model, possibly because he does not appreciate the importance of such mental mapping, possibly because he is not a clear thinker.

Section 5.2

1   If the new system will speed ordering by cutting out a clear bottleneck in the present system, it may be preferable to measure the average length of time spent by orders in the bottleneck, rather than the overall ordering-to-despatch time. However, assuming that the present system is to be tuned in a number of places and it is important to measure the overall effect, then it will be necessary to guess the standard deviation of overall order processing time. If orders currently take 4 to 40 days to complete, maybe this spans 4 to 6 standard deviations, so $s = 36 \div 4 = 9$ or $s = 36 \div 6 = 6$. The estimate of mean processing time should be fairly accurate, with accuracy limits of, say, $\frac{1}{2}$ day either way, to be reasonably sure of proper conclusions when comparing the old system with the new one. (You would want the limits of accuracy narrower than the expected 2 day saving, or the saving may be swamped out by the estimating error.) For safety's sake, taking the higher estimate of $s = 9$,

$$n = (4 \times 1.96^2 \times 9^2) \div 1^2 = 1245$$

Even the rasher estimate of $s = 6$ means 553 representative samples would be required to give an estimate of the mean to within $\frac{1}{2}$ day at the 95% confidence level.

Section 5.3

1   (a) No objective evidence – event may be long past; conducive to bias. May have forgotten; no 'can't remember' answer. Supplied answers not orthogonal, e.g. advertisement poster at school; no advice what to do.
(b) Not all that bad a question if you believe that any suggestion is better than none; but one wonders how much more effective it may be to interview selected managers. The questionnaired managers may include all those who have great ideas, but this would have no value if those managers did not respond, perhaps because they haven't the time or their answers require an essay. Maybe the question should be, 'Would you like to discuss your suggestions for improving ...'
(c) Which year? If 'stationery consumables' is an accounting head (a predefined term), why can't this information be obtained from routine accounts? If it is not, is it a term clear to the manager? If it is not clear, the answer is open to bias. If it is clear, but not an accounting head, how will the manager be able to find out the answer?
(d) Not an objective question. Most people's preferences are conditional, e.g. a pre-despatch system might be preferred **if** it didn't delay delivery, **if** it improved cash collection, **if** ... **if** ... The answers may be misleading if the conditions are not understood. Often the conditions are too various to bear analysis. Is the term 'pre-despatch' known to the sales represent-ative? The phrasing of the questions, where only one alternative is shown, may suggest the answer and induce bias in the replies. (Cf. 'Would you prefer a pre-despatch or post-despatch invoicing system ...')

REFERENCE

(1) Berdie, D. R., and Anderson, J. F., Questionnaires: design and use, Scarecrow Press, Metuchen N.J., 1974.

6 Fact recording

6.1 NCC DOCUMENTATION STANDARDS

Many standard methods of recording the facts about the existing system, or describing/defining a proposed new system, have been published. Even more have been developed by the data processing departments of corporations for internal use. Protagonists of a given method usually make some claim for increased efficacy of their particular standards; claims which are rarely backed up by objective evidence, although often plausible at face value. New methodologies tend to be tested out in the first instance by highly skilled and highly motivated workers, a combination which augurs well for success whatever the method. My own view is that successful development of a new system is not particularly sensitive to the method of documentation so it doesn't matter much which standards are adopted. Successful use of a particular methodology may even be quite sensitive to the personal characteristics of the analyst. What is important is that the analyst becomes so skilled in one particular approach that he can naturally use that approach as an aid to analysis, design and communication during development of a new system. An approach to documentation which relies upon facts being recorded after the system is operational is doomed to failure, since nine times out of ten there are other more interesting or urgent tasks for the analyst at that time.

The standards published by the National Computing Centre (1) have been adopted in whole or in part by many organisations and are probably those most widely used in practice in the UK. You are recommended to inspect a copy of these standards, since only an outline is given here.

In addition to supplying a number of useful standard forms and techniques (such as the Clerical Document Specification – see Figure 5.3), the NCC have a valuable method of organising, filing and retrieving documents collected during, or generated by, the system development process. This comprises a scheme of **files** for containing the documents, **document references** for deciding the order of the documents within the files, and **cross-references** which allow the standards to be used in a top-down approach to system description or definition.

The **files** which could be usefully maintained for development of a system are as follows.

Old system file	– containing all documents relating to the old system (if any), such as sample forms, present system clerical document specifications, grid charts and procedure descriptions;
New system file	– containing all documents relating to the current state of an implemented or proposed new system, such as reports, new system forms, new system clerical document specifications, grid charts and procedure specifications. Only the current edition of a document is maintained in this file. Rejected proposals and superseded documents are removed and placed in a

System history file.

There may be documents which relate to more than one system. Typically these go into a

Shared database file   – containing specifications of any database which is used by more than one system;

Library programs file – documentation of programs used by more than one system;

Component file   – documentation of program components (modules, subprograms) used by more than one program.

Each of these files can be conceived as containing a number of pockets. Each pocket holds a particular type of document. The **document reference** is entered on the top right-hand corner of each document. The complete reference determines the sequence of documents within the file. It is of the

Document-type

	BACKGROUND
1	Terms of reference and other background material
	COMMUNICATIONS
2.1	Discussion records
2.2	Correspondence
2.3	User manuals and sundry associated documents
	PROCEDURES
3.1	System outline or other overview such as general system flowchart
3.2	Clerical procedure flowchart, etc.
3.3	Operations department procedure flowchart, etc.
3.4	Computer run chart, etc.
3.5	Computer procedure flowchart, etc.
	DATA
4.1	Clerical document specification, sample documents
4.2	Computer file specification – source data input
4.3	Computer file specification – computer-produced documents and print layouts
4.4	Database or stored computer file specification, interactive display formats
4.5	Record layouts – source data input files
4.6	Record layouts – computer output files
4.7	Record layouts – database or stored computer files
	SUPPORTING INFORMATION
5.1	Analyses and interactions – grid charts, organisation charts
5.2	Data item definition
5.3	Hardware and software facilities
	TESTING
6.1	Test data specification
6.2	Test plans
6.3	Test operations
6.4	Test logs
	COSTS, PERFORMANCE, DOCUMENTATION CONTROL
7	Development and operation cost information
8	Estimates or reports of timings, volumes, growth
9.1	Copy control – whereabouts of report copies, etc.
9.2	Amendments incorporated list
9.3	Outstanding amendments

Fig. 6.1: Suggested NCC standard document-types for system development documents (abbreviated)

form

> system-name/document-type/document-name/sheet-number.

System-name is an analyst-invented unique name for the system, e.g. 'PRS' (short for 'Personnel Records System').

Document-type is a code used to order the documents by type within file, i.e. to identify the pocket. Standard document-types are briefly listed in Figure 6.1.

Document-name is an analyst-invented name to distinguish different documents of the same type, and to give order to documents within a type.

Sheet-number is a serial number for multi-sheet documents.

Example: PRS/4.3/MANPOWER REPORT/1 labels the first or only page of a specification of a computer-produced document in the PRS system. The specification is to be known as MANPOWER REPORT.

Cross-references may be made from one document to another by quoting the document reference. If the document referred to is in the same system, and the sheet number is irrelevant, then an abbreviated reference comprising only document-type/document-name is cited.

Questions

1   You have completed a clerical document specification for an order form used in an order entry system (OES). What reference would you put on it?

(2 min)

2   What errors are there in these references?
Organisation chart of the personnel department: PRS/1/ORG/1
Document/department grid: PRS/3.2/DOCDEP/1
Discussion record about personnel database format: PRS/4.4/DBFORMAT/1

(2 min)

3   You join an organisation and you find that they have a collection of standards for systems documentation, but these standards are not a bit like the ones you have learned. What should you do?   (2 min)

6.2 PROCEDURE DOCUMENTATION

A good target for an analyst is to confine a description of a procedure to one sheet of paper. If this cannot be done satisfactorily, then some portion or portions of the system should be **summarised** on the single sheet and each summarised portion should be **expanded** on another single sheet. This process of summarisation at the higher level and expansion at the lower level ('decomposition') can be repeated until the required level of detail is reached. Figure 6.2 shows the levels at which the NCC standard procedure-describing documents and techniques are most useful, and a possible structure for using the documents to describe or define procedures. This should not be interpreted rigidly, since the documents can be useful at other levels.

A **system outline** is illustrated in Figure 6.3. The order of listing inputs, outputs, etc. is not significant. Often there are more inputs or outputs than may be desirable to show in an overview. In this case, the less important items can be omitted and a note made of the nature of the omitted items, or perhaps they can be summarised under one term, e.g. 'control reports'.

Often, an item which is an output at one point is re-input at another point. In this case, it should appear both as an output and an input, unless it is used only for transfers of data within the system (not coming from or going to outside the system), in which case it can be omitted, or entered as a 'file', as preferred.

It is convenient to think of 'outputs' here as 'formal messages (documents) which leave the system' and inputs as 'formal messages which enter the

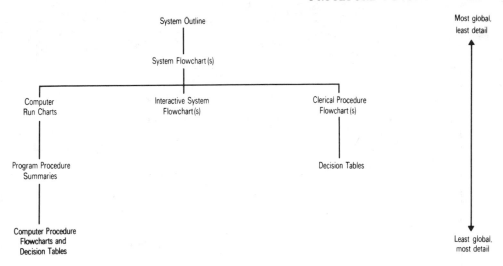

Fig. 6.2: Possible hierarchical use of standards

system, or which are created within the system from an informal message'. These are the most important things to record. Informal messages and transfer messages may be recorded at the analyst's discretion.

The system outline document is useful for recording or discussing broad details of existing or proposed systems, or as a personal aide-memoire for the analyst when he is trying to understand a system.The cross-references are likely to be entered only as more detailed documentation is completed, so that in the final form the system outline makes a complete index to procedure and data documentation.

A **system flowchart** is another overview document, organised in columnar fashion as in Figure 6.4, where each column represents a department or function. Flowchart symbols are placed in the relevant column. The slanting parallelogram represents a document, and the rectangles represent operations; but what the analyst writes in the symbols is much more important than the shape of them. The top strips in these symbols contain cross-references to data descriptions or more detailed procedure descriptions. The open arrow represents physical movement of data, not flow of the process. The box with curved sides denotes a file. The circle in the example connects flow to symbol number 1 on page 2 of the flowchart. Such cross-page connection will not be necessary if top-down decomposition is achieved as explained above.

System flowcharting is valuable for showing the flow of data and sequence of operations. Some practice is needed before skill is acquired at deploying the boxes economically and meaningfully. Although a template has been used to produce a neat example, the analyst should not be afraid to draw rough freehand flowcharts to record present processes or explore possibilities.

A **clerical procedure flowchart** may be drawn in similar fashion to a system flowchart, with each person or section heading the column (see Figure 9.5 for an example). Another approach is to chart clerical procedures in the same way as a program flowchart. This will emphasise the logical steps involved rather than the documents and people concerned (see Figures 3.1 to 3.3).

Interactive systems, computer runs and program procedures will be dealt with later.

System Outline NCC	Title BAKERY ORDER PROCESSING	System HOE	Document 3.1	Name SYS	Sheet 1

Inputs

Standing order amendments

Returns and adjustments

Payments and journal entries

Report parameters

Processes

Clerical order entry

Input validation

Update standing orders, deliveries, customer accounts

→ Produce daily schedules

Produce weekly invoices

Produce monthly statements and reports

Files

Standing orders

Customer accounts

Commodities, details and sales

←→

Outputs

↓

Delivery notes

Delivery van loading schedules

Bakery production schedules

Invoices

Statements

Sales analyses

Overdues report

Input blanks for completion (turn-around document)

Notes

See 3.1/SYSFLOW for amplification and cross-references

S 31

Author SW	Issue	
	Date	1.4 80

© 1969. The National Computing Centre Limited

Fig. 6.3: System outline

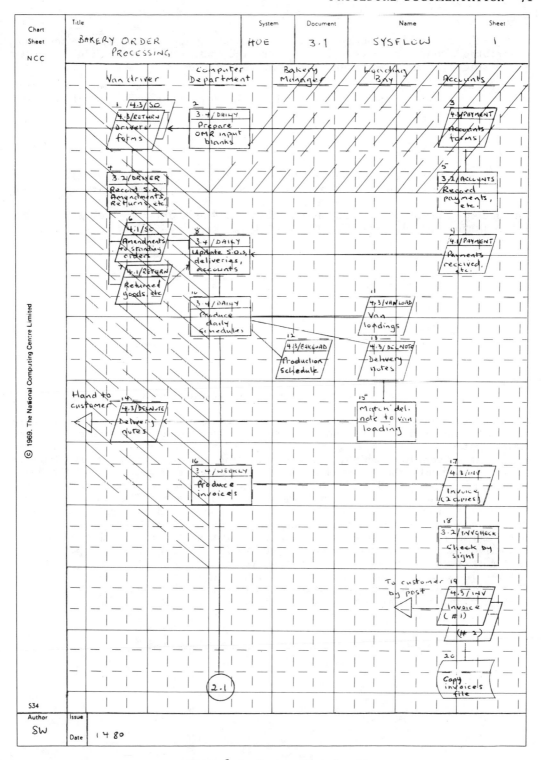

Fig. 6.4: System flowchart

Exercises

1   Draw a system outline and a system flowchart to NCC standards, describing your conceived system for the mail-order wine club's order processing system (assignment at the end of chapter 3).

2   Draw a clerical procedure flowchart for your local data preparation department's operations for receiving, punching or keying, verifying and returning data.

6.3 DECISION TABLES

A decision table can be used instead of a procedure flowchart. It is particularly useful when the alternative actions which can be taken depend upon complicated pre-conditions.

A decision table is divided into quadrants, the right half being ruled off into columns called **rules** (Figure 6.5).

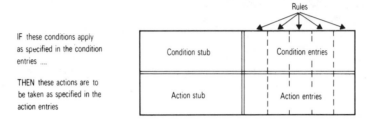

Fig. 6.5: Skeleton of a decision table

The conditions on which the actions depend are written, one per line, in the condition stub. The conditions **may** be written in any order, but are **best** written with the most pre-emptive condition first. By 'most pre-emptive' is meant a conditon which, when considering one of its outcomes (true or false), pre-empts the greatest number of later conditions, so that the later conditions do not need to be tested. Figure 6.6 shows a poorly-ordered table by this criterion and Figure 6.7 shows it well-ordered. The order in Figure 6.7 allows rules 1 and 3 of Figure 6.6 to be combined into rule 1 of Figure 6.7. The dash in rule 1 means that when the first condition is true, it is not necessary to consider the second condition. (This assumes that the conditions are to be considered consecutively starting at the top. This is the most natural view. An alternative is to comprehend in one go all the entries in a given rule. This would allow placing a dash even against the first condition. However, this usage is confusing and is best avoided.)

The actions must be sequenced in the order they are to be taken – so,

WOE/3.2/LINE BACK-ORDER/1

	1	2	3	4
Is it a widget order line?	Y	Y	N	N
Is there stock to fill the order line?	Y	N	Y	N
Fill order line	X		X	
Raise back-order		X		
Refer to customer		X		X

Fig. 6.6: 'Only widgets are back-ordered'

WOE/3.2/ LINE BACK-ORDER/1

	1	2	3
Is there stock to fill the order line ?	Y	N	N
Is it a widget order line ?	-	Y	N
Fill order line	X		
Raise back-order		X	
Refer to customer		X	X

Fig. 6.7: 'Only widgets are back-ordered' improved

in Figure 6.7, the back-order is raised **before** referring to the customer.

In a **limited-entry** table, the condition entries are only **Y** (meaning the condition is true) or **N** (false) or a dash (the condition is not to be tested in this instance). The action entries are only **X** (the action is to be taken) or blank or dash (either means the action is not to be taken). Each rule, of course, shows the actions that follow from the stipulated outcomes. Figures 6.6 and 6.7 are both limited-entry tables.

After a table has been drafted, it should be checked for completeness. It will be seen that there are four possible rules for a limited-entry table with two conditions, eight rules for a table with three conditions and, in general, 2^c rules for a table with c conditions. Each of the 2^c rules should be accounted for. Where a dash appears in the condition entries for a rule, this means that two rules have been consolidated into one; two dashes in a rule mean that four rules have been consolidated into one and, in general, d dashes in a rule means 2^d rules have been consolidated into one. So the completeness check is that 2^c = the sum of 2^d for each rule. In Figure 6.7, $2^2 = 2^1 + 2^0 + 2^0$ and in Figure 6.6, $2^3 = 2^0 + 2^0 + 2^0 + 2^0$. If 2^c is more than this sum, there is a missing rule; if less, there is a duplicate rule or an ambiguity.

Ambiguity will arise if the condition entries of two rules are the same but their action entries differ. Ambiguity or duplication is most commonly overlooked when a rule which is included in a consolidated rule also appears explicitly in its own right.

Redundancy is not an error, it means only that a consolidation opportunity has been overlooked so that the table is more elaborate than is strictly necessary. To eliminate redundancy, any rules with identical action entries should be identified and their conditions inspected to see whether any consolidation is possible.

People often get confused if they have chosen **mutually exclusive** conditions to go in the condition stub. This means, for example, that a **Y** answer to one condition automatically leads to an **N** answer to another condition. Such conditions are better rephrased so that they appear once only in the condition stub; this may entail phrasing the condition to take extended entries, as illustrated later.

To prevent error in tables, they should be kept simple, with about four conditions. More complicated cases can be dealt with by organising tables into a top-down hierarchy where an **action** in a higher table is to **perform** a lower table. Figure 6.8 gives an example. If a table is to be repeated, the control over repetition should be expressed in the performing table, not the performed table (see the last action in Figure 6.8).

The ELSE rule illustrated in the standard decision table of Figure 6.9 collects together the rules not specifically identified. It is best if each rule in the ELSE rule is analysed during construction of the table. The ELSE rule should be used only as a documentation shorthand after analysis.

Figure 6.9 also illustrates the use of **extended entries** in a table, in the condition entries for Key Comparison and T Code and the action entry for

WOE/3.2/ ORDER CHECKING/1

Is customer credit OK ?	N	Y	Y	Y
Is there a shortfall on any order line ?	-	N	Y	Y
Is it a class 'urgent' order ?	-	-	N	Y
Refer to customer	X			X
Fill order		X		
For each order line do 3.2/LINE BACK-ORDER			X	

Fig. 6.8: Top–down structured decision table

Set Error Code. An extended entry contains something other than **Y, N** or dash for a condition entry, and other than **X**, dash or blank for an action entry. The GO TO in the example is a possible use, but carries the same dangers of complexity that can arise in computer programming.

The decision table is a very powerful technique for clarifying business rules which are complex. It is also a powerful tool for definition of computer procedures, especially when used in conjunction with a decision–table pre-processor for production of programs.

Questions

1   Draw a limited–entry decision table for the following case:

'Customers who do not have a record card are not allowed credit unless the Manager gives his authorisation. Other customers are allowed credit if they do not have an outstanding balance older than three months. Credit will be refused if a shorter–term outstanding balance exceeds £10,000. Of course, none of this applies if cash is tendered with order.'

Start by identifying the condition and action phrases in the procedure.

(15 min)

2   How many rules will there be if extended entries are allowed in the condition entries?   (5 min)

3   How many rules are consolidated into one when one or more dashes appear in the rules of a table containing extended condition entries?   (5 min)

4   In Figure 6.9, what rules are included in the ELSE rule, assuming T Code can take only the values T, A, D?   (5 min)

6.4 DATA DOCUMENTATION

We have already seen an example of the NCC standard form for Clerical Document Specifications. Standard forms also exist for computer documents, visual displays, computer inputs and outputs, databases, files and records. These will be discussed later, in context.

As fact-finding and analysis progress, the analyst will find he is dealing with a growing collection of data items (synonym: data elements), such as 'customer name', 'order number'. Whatever design approach is adopted, each item will have to be named, its use identified and other characteristics documented in the specification of a new system. The NCC Data Definition (Figure 6.10) is handy for building up a **dictionary** (synonyms: directory, catalogue) of data items, which will be useful in both analysis and design stages. During fact-finding, only the top half of the form is completed, with the name of the element (PRODCODE in Figure 6.10), the source document or procedure (3.5/ASSEM), description in natural language (Product code),

Chart Sheet NCC	Title		System	Document	Name	Sheet
	Stock File Update		SOP	3.5	SUDATE	1

C = 5
A = 13
R = 14

	1	2	3	4	5	6	7	8	9	10	11	12	13	14	
End of Transaction File	Y	Y	N	N	N	N	N	N	N	N	N	N	N	N	E
End of Master Input File	Y	N	N	N	N	N	N	N	N	N	N	N	N	Y	L
Key Comparison T:MO	-	-	<	<	<	<	=	=	=	=	=	>	>	-	S
T code =	-	-	I	I	A	D	I	I	A	D	D	-	-	-	E
IND set	-	-	Y	N	-	-	Y	N	-	Y	N	Y	N	-	
Perform Update Routine	-	-	-	-	-	-	-	-	X	-	-	-	-	-	
Write MO to File	X	X	-	-	-	-	-	-	-	-	-	X	X	-	
Read new MI	-	X	-	-	-	-	-	-	-	-	X	-	X	-	
Move MI to MO	-	X	-	-	-	-	-	-	-	-	X	X	X	X	
Move T to MO	-	-	-	X	-	-	-	-	-	-	-	-	-	-	
Set IND	-	-	-	X	-	-	-	-	-	-	-	-	-	-	
Unset IND	-	-	-	-	-	-	-	-	-	X	-	X	-	-	
Read new T	-	-	-	X	-	-	-	-	X	X	X	-	-	-	
Set Error Code =	-	-	1	-	2	2	3	4	-	-	-	-	-	-	
Go to Error routine	-	-	X	-	X	X	X	X	-	-	-	-	-	-	
Go to End routine	X	-	-	-	-	-	-	-	-	-	-	-	-	X	
Go to Basic Update	-	X	-	X	-	-	-	-	X	X	X	X	X	-	
Go to Secondary Update	-	-	-	-	-	-	-	-	-	-	-	-	X	-	

Key: MI - record in master file input area
       MO - record in master file output area
       T - record in transaction file input area
       IND - indicator that MO holds previously inserted record
       A - amendment transaction
       D - deletion transaction
       I - input transaction

S34

Author	Issue	
	Date	

Fig. 6.9: Extended-entry decision table

Data Definition NCC	Product Code		(System) PROD 1	(Document) 5.2	Name PRODCODE		Sheet 1
			Source 3.5 ASSEM		Other Names PART NO SUBASS NO		
			Value Range A0001A - Z9999C				
	Category						

	Used on		Correspondence	Used by			
I/S/O	Record Name/Ref.	Picture		Program Name	R/U	Program Name	R/U
I	4.2/PROCDET	A9(4)A	→→	ASSEM 1	R		
S	4.7/PRODMAST	A9(4)A	} →→	ASSEM 2	U		
S	4.7/PRODCOMP	A9(4)A	}				

Security/Authorisation	Other Comments

R 1977 The National Computing Centre Limited

S48

Author	Issue	
	Date.	

Fig. 6.10: Data Definition

synonyms, if any (PARTNO, SUBASSNO) and range of values (A0001A–Z9999C). The system, document and category boxes appearing on the form are not used in our present context. If synonyms exist, a Data Definition should be raised for them, pointing to the main definition.

Packages exist for maintaining data dictionaries on the computer; for example, MSP's DATAMANAGER. In addition to maintaining the equivalent of manual records, such packages allow for printing out the dictionary (or parts of it) in alternative sequences, enquiries (what systems/records/programs use PRODCODE?) and for automatic generation of source program language statements defining the files and records to be used by programs.

ASSIGNMENT

Prepare a list of the data elements, their names and pictures, you envisage for the mail–order wine club's order entry system, leaving out stock replenishment and management reports (see exercise at end of section 6.2). Do not worry over much about incompleteness of your list of elements: deficiencies will later be brought to light as a result of consideration of the desired procedures.

The value of the Data Definition form or data dictionary becomes more apparent with a large system, which over a number of years grows and needs maintenance. An organised and extensible documentation of the data elements becomes an aid to productivity and helps reduce error. Using the forms for an assignment such as the above generates a large volume of paper which in the short term will aid the analyst little; this is why I ask only for a list.

In practice, system design is achieved by the analyst mentally alternating his attention between data structures and procedure definitions. It is impossible to define procedures without definition of the data elements they use and it is impossible to select the required data elements (and, often, to define their precise semantic purpose) without conception of the procedures they will support. The analyst must mentally test the internal consistency of his data/procedure definitions at every step.

The emphasis on data analysis here and in the next chapter is therefore somewhat artificial, but it prepares the ground and allows rehearsal of a desirable skill.

ANSWER POINTERS

Section 6.1

1   OES/4.1/ORDER FORM/1

2   Organisation chart, 5.1; grid chart, 5.1; discussion record, 2.1.

3   Learn to use theirs. After doing so, if you think the standards you know are better, encourage them to change.

Section 6.3

1   See Figure 6.11.

2   Let x_i be the number of extended entries possible for condition i ($x_i = 2$ in the case of a limited–entry table). The number of rules is the product of the number of possibilities for each of the c conditions, i.e.
$$\prod x_i \text{ for } i = 1, 2, \ldots, c$$

3   If x_j is the number of possible entries for dashed condition j, the number of rules consolidated in a dashed rule with d dashes is
$$\prod x_j \text{ where } j = 1, 2, \ldots, d$$

SOE/3.2/CREDIT /1

Cash with order ?	Y	N	N	N
Customer has record card ?	-	Y	N	N
Does manager authorise credit ?	-	-	Y	N
Allow credit			X	
Refuse credit				X
Perform 3.2/CREDIT DETAIL		X		

SOE/3.2/CREDIT DETAIL /1

Account balance older than 3 months ?	Y	N	N
Account balance exceeds £10 000 ?	-	Y	N
Allow credit			X
Refuse credit	X	X	

Fig. 6.11: Solution to question 1, section 6.3

4   None. Presumably this mistake arose out of a desire to illustrate an ELSE rule.

(Number of rules = $\prod x_i$ = 2 x 2 x 3 x 3 x 2 = 72

Number of rules accounted for = $\sum \prod x_j$ =

Rules 1, 2, 14: 3 x 3 x 2 = 18	Total for 3 rules = 54
Rules 12, 13: 3	Total for 2 rules = 6
Rules 5, 6, 9: 2	Total for 3 rules = 6
Rules 3, 4, 7, 8, 10, 11: 1	Total for 6 rules = 6
	Total = 72)

REFERENCES

(1) National Computing Centre, **Data Processing Documentation Standards,** NCC Publications, Manchester, England, 1977.
(2) Installing a data dictionary, **EDP Analyzer, 16,** 1, January 1978.

7 Data analysis

7.1 IMPLEMENTATION-INDEPENDENT DATA ANALYSIS

Data analysis means here the identification of the data elements which are needed to support the data processing system of the organisation, the placing of these elements into logical groups and the definition of the relationships between the resulting groups. No standard terminology for data analysis has emerged. Following some of the terms used in this chapter, synonymous terms used in some other works are shown in order to help the reader to tie together the ideas of different authors should he desire.

  Systems analysts often, in practice, go directly from fact-finding to implementation-dependent data analysis. Their assumptions about the usage of, properties of, and relationships between data elements are embodied directly in record and file designs and computer procedure specifications. The introduction of Data Base Management Systems has encouraged a higher level of analysis, where the data elements are defined by a logical model or 'schema' (synonyms: conceptual schema, symbolic model). Often debate about the schema in the context of a Data Base Management System involves consideration of the effect of alternative designs on the efficiency or ease of implementation. In other words, the analysis is still somewhat dependent on the implementation. If we consider the data relationships, usages and properties that are important to the business without regard to their representation in a particular computerised system using particular software, we have the present subject, implementation-independent data analysis.

  It is fair to ask why data analysis should be done if it is possible, in practice, to go straight to a computerised system design. Data analysis is time-consuming; it throws up a lot of questions. Implementation may be slowed down while the answers are sought. It is more expedient to have an experienced analyst 'get on with the job' and come up with a design straight away. The main difference is that data analysis is more likely to result in a design which meets both present and future requirements, being more easily adapted to changes in the business or in the computing equipment. It can also be argued that it tends to ensure that policy questions concerning the organisation's data are answered by the managers of the organisation, not by the systems analysts. By contrast, the expedient approach is often vulnerable to changed requirements or equipment because there may be unjustified assumptions embodied in the computer system design. Management are more likely to find they have lost control over data storage policies. Data analysis may be thought of as the 'slow and careful' approach, whereas omitting this step is 'quick and dirty'.

  From another viewpoint, data analysis provides useful insights for general design principles which will benefit the trainee analyst even if he finally settles for a 'quick and dirty' solution.

  Data analysis is not an easy subject to explain or to learn. It is a short cut to a position of understanding which for most analysts is reached only after years of experience, if then. You may have difficulty understanding this chapter until you have mastered the link between it and chapters 11 and 12, and tried out the ideas in practice.

Question

1   Is 'slow and careful' generally to be preferred to 'quick and dirty'?

(10 min)

7.2 ENTITIES, RELATIONSHIPS AND ROLES

Recall that a formal message is one whose type of content is predefined. The **ideal** objective of data analysis is to build a model of the data which supports present needs for formal messages together with all future formal message needs of the organisation. Such an **ideal** model (synonyms: conceptual data model, best model, mental model) is unlikely in practice to be achievable, so the analyst has to seek a compromise between meeting possible future needs and delaying implementation for prolonged fact-finding.

This practical ideal model can be diagrammed by representing entities as boxes, and relationships as diamonds, as in Figure 7.1.

Fig. 7.1: Entity-relationship diagram

An **entity-type** (synonym: entity-set) is a class of object, person, event, place, operation, or other feature which exists in the object system and which is to be the concern of a formal message. An **entity-occurrence** is an instance of an entity-type. CUSTOMER(NAME, ADDRESS) is an entity-type; (A. Jones & Son, High Street Newtown) is an occurrence of CUSTOMER. PRODUCT(DESCRIPTION, COLOUR, PART #) is an entity-type; (widget, yellow, 11747) is an occurrence of the PRODUCT entity-type.

It is open to the analyst to choose entity-types in any way which he believes will be meaningful in the model. A poor initial choice can be corrected by subsequent analysis, so there is no need to delay making the decisions. A good starting point generally is to choose as entity-types those features of a system which are fairly stable and persistent; the sort of thing about which it is likely to be worthwhile to have a permanent record which will be kept up-to-date or preserved for reference.

A **relationship-type** (synonym: relationship-set) is a class of event, place, operation, transaction or other feature of the object system which connects two or more entities and which is to be the concern of a formal message. There is no point in recording other relationships. A **relationship-occurrence** is an instance of a relationship-type. In Figure 7.1, CUSTOMER ORDERS PRODUCT is a relationship-type; an order instance such as (number 11760 placed by A. Jones & Son on 1st March for 100 widgets and 50 gadgets) is an occurrence of this relationship-type. All relationship-types which are necessary to support the formal messages should be identified. Entity-occurrences may enjoy the relationship with one another in one-to-one fashion, or one entity-occurrence on one side may be related to several on the other side (a one-to-many relationship). If an entity-occurrence on either side may enjoy a given relationship-type with several on the other side, the relationship is said to be many-to-many.

It is usually straightforward to name the chosen entity-types using the terminology current in the organisation. Naming relationship-types sometimes poses more problems. In Figure 7.1, the use of the verb wrongly introduces an idea of direction into the entity-relationship diagram, and the data which is to be stored about the relationship may concern a time other than the day of ordering (for example, the amount of the goods delivered in response to the order). The best convention is to name the relationship with the joined entities, e.g. CUSTOMER–PRODUCT, or, if a meaningful word can be found, a noun, e.g. ORDER. Thus, ORDER is a relationship which joins customer and product. There is no sense of flow in an entity–relationship diagram. It just displays the chosen entities and some of the permanent or temporary relationships which exist between them.

People sometimes choose documents in the existing system as entities or relationships. This is not intrinsically wrong, but there is often a simpler view which can be taken by asking oneself what those documents are about. For example, an existing system may have an order form, a picking note, a despatch note and an invoice, but it is not usual to see these as four entities or relationships. All those documents are passing messages about a particular transaction; it is that transaction, which could be named ORDER, which needs to be modelled in the entity–relationship diagram.

It is a good idea to keep the number of entities and relationships down (Figure 7.10 illustrates a realistic case).

It is possible for two entities to enjoy more than one relationship–type between them. For example, customers may order products and also complain about products. Figure 7.2 illustrates how a **role** can be introduced to qualify an entity when it engages in different types of relationship, should it be deemed necessary to model them all.

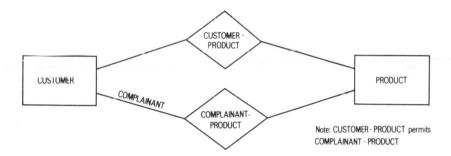

Fig. 7.2: The role COMPLAINANT is a sub–type of CUSTOMER; see text for explanation of note

It should not be inferred from these examples that CUSTOMER, PRODUCT etc. are invariably entities in an order processing system. The desired entities are those which are needed to support the data processing system of the organisation and which will be widely recognised, or endorsed as fitting, by the managers of the organisation. The entities will therefore be dictated not only by the type of business but also by the norms of the people in it. A wholesale bakery may see CUSTOMERs and PRODUCTs, but a grocery supermarket may see SALEs and LINEs. Although a supermarket does have customers, an individual customer is not normally the concern of a formal message. A life insurance company may see POLICYHOLDERs, LIFE–ASSUREDs and TYPE–OF–INSURANCEs; a shipping company may see CONSIGNORs, CONSIGNEEs and CONSIGNMENTs.

Existence–dependence arises when one entity–occurrence depends for its existence on another entity–occurrence. If order–line is existence–dependent on order then if an order is cancelled the related order–lines are also to be cancelled. The special features of a relationship, such as existence-dependence, one relationship excluding another, one relationship permitting another, should be noted if they are important to preserving the consistency of data stored in accordance with the model. These special features can be used later to specify computer controls over the data accepted or stored, and to assist the analyst in evaluating alternative designs of the computer files and records.

Questions

1   In Figure 7.3, what do you think is intended by the ASSEMBLY relationship?   (2 min)

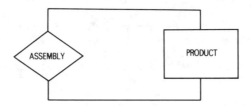

Fig. 7.3: Question 1, section 7.2

2   Without worrying too much over complications, what would be a suitable draft entity-relationship diagram for the reserving, loaning, fining-for-overdue and acquiring-new-stock systems for a public lending library? Loans are to be made from the loan stock of books, reservations are accepted on titles which are in stock (more than one copy of a title may be stocked) or which are the subject of a purchase order. Only one title is ordered on a purchase order.   (20 min)

7.3 ATTRIBUTES AND VALUES

An **attribute** is a class of number, name, value, quantity, place, time, other feature of an entity-type or relationship-type, whose value is to be contained in (or is otherwise some determinant of) a formal message. For example, a CUSTOMER entity-type may have the attributes CUSTOMER-NO, CUSTOMER-NAME, CUSTOMER-ADDRESS, CUSTOMER-POSTCODE, CUSTOMER-CREDIT-LIMIT, etc. It is a common convention to record the attributes in brackets after the entity-type name or relationship-type name. The ellipsis (...) is used to show there may be other attributes needed in practice which are omitted in the example. The attribute-names should be the same as those recorded on the Data Definitions (see section 6.4). The **values** of attributes of a particular occurrence of the CUSTOMER in Figure 7.4 could be, for example, (1230, A. Jones & Sons, High Street Newtown, ...).

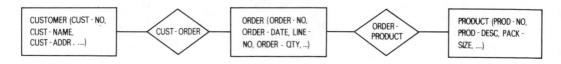

Fig. 7.4: Entities and attributes (before normalisation, discussed later)

A relationship always has, as attributes, at least the 'key' attributes (described in more detail later in this section) of the entities it joins. Other than these, no attribute should appear more than once. The attributes that the analyst identifies as being concerned in messages must all be accounted for eventually. They will either belong to the entities or relationships, or not belong to any conceived entity or relationship. Attributes legitimately in this last category are typically parameters of the system – such as the current date or time, current tax rates, current legal values of a code, the next customer number to allocate – or attributes which are incidental to messages or derived from other attribute values, such as the page number of reports, a report total or control total of a particular attribute in all occurrences. The analyst is best advised to start by concentrating on the attributes that belong to entities and relationships; the other attributes can be brought into account later when business and control procedures are established. Entities and relationships are like the logical records defined in computer files, but to call them records would be presumptive at this stage. Records will be defined for the computer system and may, or may not, be equivalent to the entities and relationships conceived at this time for the data model.

It is often convenient to think of one of the possible values of an attribute as being a 'null' value. To say an attribute has the value NULL is to say that this attribute does not exist for, or is irrelevant to, the particular entity- or relationship-occurrence in question. Alternatively, it could mean that the attribute value has not been recorded for the particular entity-occurrence. If CUSTOMER-POSTCODE of a particular entity has the value NULL, this means either that the particular customer concerned does not have a postcode, or that he has a postcode but it is not on record. One of the analyst's objectives in system definition is to define or list the values which may be taken by each attribute (called the 'value domain'), so that checks can be made on the validity of the data recorded. Another is to consider what action is required if a NULL value is encountered.

Now that we have defined entities, relationships, attributes and values, we can redefine a **formal message** as a message or record, concerning one or more entities or relationships, whose content, as regards the attributes and constants contained in or otherwise determining the message, is completely predefined. Where the predefinition allows for alternative attributes or constants, these must be enumerated or some rule must be supplied which identifies the possible alternatives.

Questions

1   Continuing question 2 of section 7.2, what attributes would you identify for your entities and relationships?   (20 min)

2   Can an entity or relationship have no attributes apart from the 'key' attributes?   (3 min)

3   What is the real difference between an entity-type, relationship-type and an attribute? Or is there none?   (5 min)

4   What would you say to someone who asked you to draw up an entity-relationship diagram for a system to retrieve information for researchers using a general-purpose reference library?   (5 min)

5   In Figure 7.4, do you think that PACK-SIZE could be an attribute of ORDER rather than PRODUCT in some systems? Explain.   (5 min)

7.4 NORMALISATION OF RELATIONS

The analyst should aim for a **fully normalised** entity-relationship diagram. This means, loosely speaking, that the chosen entities and relationships are the **most primitive** entities and relationships. It means there will be a good place for each attribute and each attribute will go into its one proper place. A fully-normalised entity or relationship, to an analyst, is rather like an atom or atomic bond to a chemist. The chemist can better understand observed molecules, or even speculate about hypothetical molecules, when he has knowledge of the atoms and bonds concerned. An analyst can better understand an existing system of records, or speculate about hypothetical systems of records, when he has knowledge of primitive entities and relationships.

To use another analogy, learning to construct a fully-normalised entity-relationship model is like learning to use a natural language. If one tries to learn a natural language by starting off with a definition of 'a word', one is drawn along a rather tortuous philosophical thread. Yet most people can pick up a language by hearing or seeing enough examples. Even after having learned the language, though, they may still be unable to give a definition of 'a word'.

This is to explain that in this chapter I am trying only to get across the idea of full normalisation in the hope that through this discussion and later examples the analyst will learn to recognise a fully-normalised entity when he sees one. In furthering this intention, I may take certain liberties with precision. This is not to say that a more rigorous treatment is not a valuable thing to follow: an understanding of the theory of normalisation can help an analyst with difficult cases in the same way as a knowledge of grammar or etymology can help with difficulties in language. Date (1) and Deen (2) have the more precise arguments.

A **relation** (do not confuse with relation**ship**) is the name given to the collected occurrences of a particular entity-type or relationship-type. The occurrences of an entity or relationship, to a relation, are what the records are to a file in computer terminology. A relation can be illustrated as a table, as shown in Figure 7.5. Each row of this relation is called a **tuple**

PRODUCT entity-relation	PROD-NO	PROD-DESC	PACK-SIZE	QTY-IN-STOCK
	0001	HAMMER	1	127
			12	10
			144	2
	0002	CHISEL	1	46
	1147	WIDGET	1	50
			100	0
(Alternative representation)	0001	HAMMER	1, 12, 144	127, 10, 2
	0002	CHISEL	1	46
	1147	WIDGET	1, 100	50, 0

Fig. 7.5: Sample data in an entity-relation

of values. The **primary key** of a relation is an attribute (or more than one attribute) singled out because it uniquely identifies a tuple (that is, it identifies one particular occurrence of the entity or relationship and no other). In Figure 7.5, PROD-NO is a promising candidate to fulfill this role.

A relation which has null values in the primary key of some tuples (or, to take an alternative view, a relation which has multiple values for one attribute of a single occurrence of the entity), as in Figure 7.5, is said to be **unnormalised**. The operations involved in **normalisation** allow the analyst to explore alternative representations of the entities or relationships. A relation which is not amenable to further normalisation is **fully normalised** (synonyms: in fourth normal form, 4NF).

One operation of normalisation makes sure that there are no NULL values of the primary key (or ensures that multiple values for an attribute of an entity are eliminated). The following alternatives are possible:

1   redefine the attribute in question as an array-attribute;
2   partition the attribute in question into different attributes;
3   remove the attribute in question from this relation and associate it with a different or new entity-type or relationship-type (any new entities made

here will have a many-to-one relationship with the original entities;

4   Consider each value of the attribute in question as giving rise to a new entity-occurrence (we will see later that this effectively redefines the nature of the entity-type).

We now consider these possibilities with respect to Figure 7.5.

1   We could try to consider PACK-SIZE as having an array of values. But there are clues to suggest that the array idea is ungainly in this case. There would be a variable number of elements in the array or there would (commonly?) be some elements, or all but one element, which have no value; there is no clear maximum to the array; there is no clear relationship between elements which singles out the meaning of each element (they would simply be a string of elements with no way of identifying a particular one other than arbitrarily). A case of an array attribute would be as in the relation STANDING-ORDER (CUST #, TYPE-OF-BREAD, NO-OF-LOAVES(6)), where NO-OF-LOAVES has six values, one for each working day of the week. The subscript of the array has an implicit meaning – 'the day of the week' – so it is as if each NO-OF-LOAVES has a separate name, and the rules for formulating the name (of a value with a known semantic purpose) are understood. This is a special case of partitioning, below.

2   We could try to partition PACK-SIZE into, say, SMALLEST-SIZE, LARGEST-SIZE, INTERMEDIATE-SIZE. But do we know there is a maximum of three values? Also, this partitioning, as with the array of fixed size, presumably gives rise to a large number of null values. Things would be different if we knew it was a **policy** of the business to supply each product in exactly three alternative pack sizes (not necessarily the same three alternatives for different products). Then partitioning would seem to fit the facts. Maybe something like this would arise in the case of a wall-tile manufacturer who produces tiles which will only ever be of one or two colours: PRODUCT (TILE-NO, COLOUR), where COLOUR was a string of one or two values, might be normalised as PRODUCT (TILE-NO, BACKGROUND-COLOUR, PATTERN-COLOUR). PATTERN-COLOUR would be null in the case of a plain tile. PACK-SIZE does not seem to be a case like this.

3   We could certainly remove PACK-SIZE and make it a new entity with only one attribute: PACKSIZE (PACK-SIZE). This means we would have to consider a new relationship, the PRODUCT-PACKSIZE relationship. Creating new entities from a singleton attribute in an existing entity often (but not always) only complicates the data analysis unnecessarily. It would be likely to be the most fitting approach in this case particularly if PACK-SIZE is a focus of attention or a useful piece of data in its separate state, independent of the products (e.g. people are to ask questions like 'How many different pack sizes do we have?' or people want to record a pack size even when no products as yet come in that size). It is usually valuable to create a new entity only when there is more than one attribute to be removed in order to achieve normalisation, and those attributes are related (described in more detail later).

4   The last alternative is always a useful starting point even if it fails to achieve full normalisation. This is to consider each tuple, that is each fresh occurrence of PACK-SIZE within PROD-NO, as giving rise to a new entity; Figure 7.6 overleaf. In order to identify a tuple uniquely, PACK-SIZE must contribute to the primary key: PRODUCT (PROD-NO, PACK-SIZE, PROD-DESC, QTY-IN-STOCK, ...). In effect, this normalisation has redefined the PRODUCT entity-type, as it was originally conceived, into a new 'PRODUCT-with-PACK-SIZE' entity-type.

But although this example is normalised, it is not yet **fully** normalised. This is because within the PRODUCT-with-PACK-SIZE entity there exists another, smaller unit which is also identifiable as an entity – the product itself.

PRODUCT entity-relation

PROD-NO	PROD-DESC	PACK-SIZE	QTY-IN-STOCK
0001	HAMMER	1	127
0001	HAMMER	12	10
0001	HAMMER	144	2
0002	CHISEL	1	46
1147	WIDGET	1	50
1147	WIDGET	100	0

Fig. 7.6: The example of Figure 7.5, normalised

PROD-DESC is an attribute of the product (we assume), not of the PRODUCT-with-PACK-SIZE. PROD-NO is said to **determine** PROD-DESC. On the other hand, QTY-IN-STOCK **is** an attribute of the PRODUCT-with-PACK-SIZE entity; it is PROD-NO and PACK-SIZE together which determine QTY-IN-STOCK. To achieve full normalisation, we can concieve that there are three primitive components concerned: a PRODUCT entity with attributes (PROD-NO, PROD-DESC); a PACKSIZE entity with one attribute only (PACK-SIZE); and a relationship between them, PRODUCT-PACKSIZE, with attributes (PROD-NO, PACK-SIZE, QTY-IN-STOCK). The **acid test** of **full** normalisation is that **every determinant of an attribute in a normalised relation is also potentially the primary key of the relation.** In our view of PRODUCT-with-PACK-SIZE (Figure 7.6), PROD-NO alone was a determinant of PROD-DESC but PROD-NO alone was not sufficient to be the primary key of the whole relation.

It will be seen that this analysis has led us back to possibility 3 above, PACKSIZE being a separate entity. If it is the case that PACKSIZE is **not** of special interest in its own right, but is of interest only because of the relationship it enjoys with PRODUCT, no separate record of PACKSIZE will be called for in the eventual computer system design; a record of the relationships will be sufficient. This aspect is discussed further in chapter 11.

With practice, the analyst is likely to conceive fully-normalised entities and relationships in the first instance. However, in order to illustrate normalisation operations further, let us take a more substantial, but perhaps artificial, example. Consider the unnormalised entity-relation SOLDIER (NUMBER, NAME, COURSE-TITLE, COURSE-CODE, COURSE-DATE, SESSION-NO, GRADE) with sample data as in Figure 7.7.

Soldiers take courses on certain dates and get grades. The first time a course is run it is given a SESSION-NO of unity; subsequent runs of it increase this SESSION-NO by unity. Sometimes, as in the case of Jones, a soldier has to repeat a course.

To achieve normalisation here, perhaps it is immediately apparent that NAME is determined only by NUMBER and we could consider an entity SOLDIER (NUMBER, NAME), to be separated out. We would have to preserve the key of this entity in the remaining relation if we are not to lose any information. The remaining relation would be normalised if COURSE-CODE and SESSION-NO

SOLDIER

NUMBER	NAME	COURSE-TITLE	COURSE-CODE	SESSION-NO	COURSE-DATE	GRADE
1110	SMITH	BASIC TRNG	001	23	1.1.80	A
		ELEC ENG	123	3	1.4.80	A
		GUNNERY	427	23	1.5.80	B
1198	JONES	BASIC TRNG	001	23	1.1.80	B
		ELEC ENG	123	3	1.4.80	F
				4	1.9.80	D
2744	SMITH	BASIC TRNG	001	24	1.5.80	A

Fig. 7.7: Unnormalised relation SOLDIER

were to participate with NUMBER in the primary key; this would give the normalised relation, say, SOLDIER-STUDY-SESSION (NUMBER, COURSE-CODE, SESSION-NO, COURSE-TITLE, COURSE-DATE, GRADE). SOLDIER as now defined is fully normalised, but SOLDIER-STUDY-SESSION is not. NUMBER determines a single NAME occurrence, NUMBER/COURSE-CODE/SESSION-NO determines a single occurrence of COURSE-TITLE/COURSE-DATE/GRADE. However, COURSE-TITLE depends **only** on the COURSE-CODE attribute of the primary key of SOLDIER-STUDY-SESSION, while COURSE-DATE depends **only** on COURSE-CODE/SESSION-NO. Removing to a separate relation those attributes determined by only a part of the primary key leaves SOLDIER-STUDY-SESSION (NUMBER, COURSE-CODE, SESSION-NO, GRADE) which connects a SOLDIER to, say, a COURSE-SESSION (COURSE-CODE, SESSION-NO, COURSE-TITLE, COURSE-DATE). This last relation is also eligible for the same operation of normalisation, since COURSE-TITLE is determined by COURSE-CODE alone. We may conceive the entity COURSE-SESSION (COURSE-CODE, SESSION-NO, COURSE-DATE) which is connected to, say, COURSE (COURSE-CODE, COURSE-TITLE) by a relationship which is concerned only with the primary keys, COURSE-CODE and SESSION-NO. These relations now defined are fully normalised.

In this particular example there was a choice of primary key after the first normalisation operation; let us explore this alternative to see where it leads. Assuming a soldier may follow only one course at a time, we could have defined SOLDIER-STUDY-SESSION (NUMBER, COURSE-DATE, COURSE-CODE, SESSION-NO, COURSE-TITLE, GRADE). Given a NUMBER and a COURSE-DATE, we could find out the relevant values of COURSE-CODE, SESSION-NO, COURSE-TITLE and GRADE. It is not possible to apply quite the same operation of normalisation to this version, because COURSE-CODE and SESSION-NO are not dependent on COURSE-DATE (more than one course-session can start on a given date), whereas previously it was true that COURSE-DATE was dependent on COURSE-CODE and SESSION-NO (a given course-session can start on only one date). However, another normalisation operation can come into play, which seeks to ensure that the relation is free of interdependence between attributes which are **not** part of the primary key. If any non-primary-key attributes determine the value of another attribute, as with COURSE-CODE which determines the value of COURSE-TITLE, these attributes should be split off into another relation. In this example we create a new entity-relation COURSE (COURSE-CODE, COURSE-TITLE), which would leave SOLDIER-STUDY-SESSION (NUMBER, COURSE-DATE, COURSE-CODE, SESSION-NO, GRADE). The former is clearly fully normalised, but does the latter pass the acid test? The answer is **no**; COURSE-CODE and SESSION-NO together determine COURSE-DATE, but they are not sufficient alone to be the primary key of the relation. We are obliged once more to admit the existence of an entity called COURSE-SESSION (COURSE-CODE, SESSION-NO, COURSE-DATE), and once more if we remove this entity we are left with SOLDIER-STUDY-SESSION (NUMBER, COURSE-CODE, SESSION-NO, GRADE).

Questions

1   Revise Figure 7.4 so that the entities and relationships are fully normalised.   (3 min)

2   Were your entities for question 1 of section 7.3 fully normalised? (10 min)

7.5 CHECKING OUT THE DATA ANALYSIS

The entity-relationship model should be checked out to ensure that each attribute appears in only one place, except that a relationship must have, somewhere among its attributes, the primary key attributes of the entities it joins. These attributes may, or may not, constitute the primary key of the relationship; if they do, it follows that the relationship, for the purpose of the system of formal messages, can be enjoyed only once between particular entity-occurrences. For example, if CUSTOMER-PRODUCT describes a relationship in which customers order one product at a time, then CUSTOMER-PRODUCT(CUST #, PROD #, ...) would mean that only one instance of a particular customer ordering a particular product will exist, or that all the formal messages can be produced by keeping a record of only one such instance (e.g. the latest instance). If a customer may have two separate orders for a given product, and it will be necessary (as it usually will in this case) to allow both instances to exist in the database at the same time, it will be necessary to distinguish between these different occurrences of the relationship, e.g. CUSTOMER-PRODUCT(ORDER #, CUST #, PROD #, ...).

The analyst's attempts to model the data typically give rise to a number of questions (e.g. from section 7.2, can a borrower be fined more than once in connection with a given loan?). Most of these questions can be answered only by user management. The analyst should systematically record all those questions which are not answered by his earlier fact-finding and check out his assumptions. The answers should be widely agreed by the users.

The analyst can also ensure that his model will stand up to the needs of the organisation by checking out the following features: optionality of attributes; optionality and degree of relationships; responsibilities and authorities.

A list of attributes for a fully-normalised entity-type or relationship-type, as in SOLDIER(NUMBER, NAME), can be translated into a set of assertions:

1   every soldier is to have one and only one number;
2   every soldier is to have at most one name, but this is optional.

The first assertion is made because NUMBER is a primary key attribute. The second is made because a null value has not been explicitly prohibited. If consideration of the usage of the attributes leads to the conclusion that a null value is not admissable – in this case, the user management do not want a SOLDIER to be recorded with a NUMBER but without a NAME – then a mandatory form of assertion must be made:

2   every soldier is to have one and only one name.

As a result of consideration of this assertion, perhaps it may be decided to partition NAME into, say, LAST-NAME and OTHER-NAMES. Leaving aside this aspect to preserve simplicity of the example, it may be that an **alternative** name is in contemplation. Maybe a soldier can change his name, or a female soldier acquires a new name on marriage. If it were decided that old and new names may have to be recorded, the attribute list could be extended to SOLDIER(NUMBER, NAME, PREV-NAME) and the assertions to test are:

1   every soldier is to have one and only one number;
2   every soldier is to have one and only one name;
3   every soldier is to have at most one previous name.

To the extent that it has not already been obtained in previous fact-finding, approval of these assertions should be the prerogative of user management, not the systems analyst. Any list of assertions should be concluded with:

4   Bearing in mind the attributes declared for the other entities and relationships, these are the only attributes of interest for SOLDIER.

Another example. SOLDIER-STUDY-SESSION in section 7.4 leads to the assertions:

1   every soldier-study-session has one and only one (soldier-) number and course-code and session-no;

2   every soldier–study–session has at most one grade;
3   these are the only attributes of interest for soldier–study–session.

The degree and optionality of a relationship can be analysed on a draft entity relationship diagram by placing on each line between an entity and a relationship a pair of digits or values in the form (a)b; see Figure 7.8.

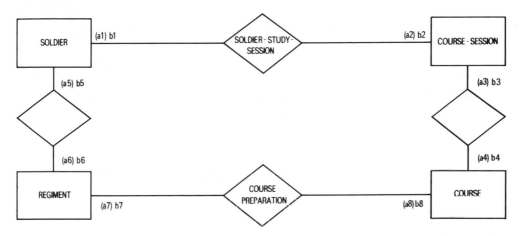

'Fig. 7.8: General form for recording optionality and degree. (The regiment–course relationship arises because a particular regiment prepares, and gives, a particular course.)

The first digit (a) shows the **minimum** number of **times** an entity may participate in the relationship; generally 0 if participation is optional, 1 if participation is compulsory. In Figure 7.8, if it is possible for some soldier not to have been on any course-session, a1 = 0. If a course-session can be planned without having a soldier on it, a2 = 0. If a course-session cannot be planned unless the course is on record, then a3 = 1, i.e. every course-session must participate in the relationship with at least one course. If a course can be planned even though it does not give rise to a course-session, a4 = 0. If a soldier must belong to a regiment, a5 = 1; if a regiment cannot exist without at least one soldier, a6 = 1.
The second digit shows the **maximum** number of **entity-occurrences** which may participate in the relationship with a single entity on the other side. If there is no definite maximum, this can be stated simply as 'many' or N. Thus b1 = many (many soldiers may go on a single course-session) and b2 = many (many course-sessions may be attended by a single soldier). Several course-sessions may stem from a single course (b3 = many) but only one course may give rise to a particular session (b4 = 1). You may care to continue in this vein and compare your answer with Figure 7.9 overleaf.
An assumption of a many-to-many relationship is slightly safer since assumptions of lesser degree may lead to limitations when embodied in a computer system design. On the other hand, unnecessary many-to-many relationships may be costly in terms of computing resources. A knowledge of optionality will be important in preserving the internal consistency of the computer system. Optionality and degree of relationships should therefore be checked out.
Assertions to test can be derived as follows for an entity-1 which is (a)b related to entity-2:

1   if a = 0, assert entity-1 may exist without an entity-2, else assert that entity-1 may not exist without an entity-2;
2   assert that (only) b entity-1's can enjoy the relationship with a given

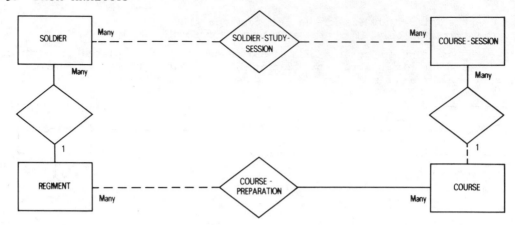

Fig. 7.9: Assumed optionality and degree. (In this polished version of the entity-relationship diagram, a=0 is shown by a dashed line and a=1 is shown by a solid line.)

occurrence of entity-2.

Applying this to Figure 7.9, the assertions to test are:

1   a soldier may exist without a course-session;
2   a course session may exist without a soldier;
3   many soldiers may attend a particular course-session;
4   many course-sessions may be attended by a particular soldier;
5   a course-session may not exist without a course;
6   a course may exist without a course-session;
7   many course-sessions may stem from a particular course;
8   only one course may give rise to a particular course-session;
etc.

Mutually exclusive relationships, permitting relationships etc. should also be checked out widely with users since they are also important to internal consistency of the database.

The remaining questions concern responsibilities and authorities for data. Entity- or relationship-occurrences may be added to or deleted from the database. Responsibilities for such creation/deletion should be ascertained. Attributes may be initially recorded, subsequently modified or inspected. Responsibilities for recording and modifying the values of attributes, and authorities for inspecting them, should be ascertained.

Armed with the data analysis, the analyst is in a position to ensure that the business procedures to be followed by people and machines in the new system conform to the assumptions, thereby reflecting the policy of management as regards data collection and control.

Questions

1   Complete the assertions above (number 9 onwards) about optionality and degree for Figure 7.9.   (5 min)

2   The Data Definition form (section 6.4) can be used to document entities and relationships as well as data items (attributes). What additional boxes would you add to the standard form to record responsibilities and authorities?   (5 min)

HOE/3.1/DATA MODEL/1

TRANSACTION (TRANS #, TRANS-CLASS, TRANS-DATE, AMOUNT)
CUSTOMER (CUST #, CUST-CLASS, BREAD-DISC, CONFECTIONERY-DISC, INV-
          NAME-ADDR, CONS-NAME-ADDR, CURRENT-BALANCE, SIGN-OF-BAL,
          (PREVIOUS-BALANCE, SIGN-OF-PREV-BAL) (3))
STANDING ORDER LINE (CUST #, COMM #, REGULAR-REQUIRED-QUANTITY (6),
          REQUIRED-QUANTITY-THIS-WEEK (6), QUANTITY-DELIVERED-THIS-WEEK (6))
COMMODITY (COMM #, PACK-QUANTITY, PACK-PRICE, PACK-POINTS,
          COMMODITY-DESCRIPTION, (WEEKLY-SALES (8), MONTHLY-SALES (24))(24))
TRANS-CUST (CUST #, TRANS #)

Fig. 7.10: The entity–relationship diagram of a wholesale bakery order processing system

ASSIGNMENTS

1   Get hold of a case study where computer files and records are described (see Figure 4.1). Construct the data model which you think the analyst had in mind. Normalise this model.

2   Prepare an entity–relationship diagram for the mail-order wine club's order processing system (see assignment at end of chapter 6). Make reasonable assumptions and prepare a checklist of questions for the management of the club. List the attributes of the entity–types and relationship–types, showing primary key, as in the examples in this chapter. Aim for a simple scheme and try to restrict the number of entities by concentrating on those that you are confident will be needed.

ANSWER POINTERS

Section 7.1

1   It is reasonable to suppose that the answer to this question lies partly with the expectations of the sponsor. If he sees the system as fulfilling a long–term need, if he believes in planning for the distant future, and if he expects the investment to be repaid over a large number of years (more than five, say), then he may prefer 'slow and careful' to 'quick and dirty'.

If the sponsor expects his investment to be repaid over a small number of years (say, less than three), and if he believes that the longer–term future must look after itself and be the subject of fresh short–term decision-taking when the time comes, then 'quick and dirty' is likely to be preferred to 'slow and careful' (this assumes that both approaches have equal chance of success in the short term).

On the whole, sponsors in large, established, stable organisations are perhaps more likely to take the first view. Sponsors in small, new organisations which are more at the mercy of market pressures are perhaps more likely to take the second view.

The contradiction to this is that the systems developed by the data analysis approach are supposed to have an increased chance of surviving change.

One can argue that a quick and dirty solution is likely to last a longer time in a stable organisation than in one which has many changes, at any rate if stability of the organisation is matched by stable data processing system requirements.

Maybe the simplest short answer is that where planning horizons are distant and system change will be frequent, 'slow and careful' is indicated. Where planning horizons are near and system change will be infrequent (e.g. the system has a short life), 'quick and dirty' is indicated. In-between cases are not clear-cut; 'quick and dirty' is more risky, 'slow and careful' is more costly initially.

Section 7.2

1   A given entity-occurrence of type PRODUCT may be assembled from other entity-occurrences of type PRODUCT. 'Standard lamp' may be a product, assembled from a lamp-base product and a lamp-shade product.

2

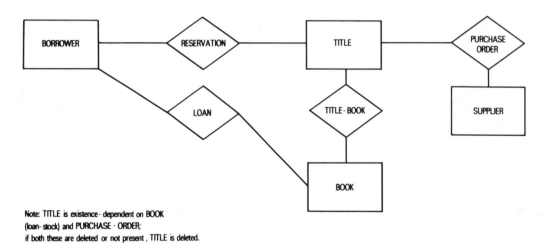

Note: TITLE is existence-dependent on BOOK
(loan-stock) and PURCHASE-ORDER;
if both these are deleted or not present, TITLE is deleted.

Fig. 7.11: Possible answer to question 2, section 7.2

This answer assumes that a fine is made only once on a given loan. If a borrower may be fined several times in connection with a single loan, a FINE could be conceived as a separate entity also participating in the LOAN relationship.

Section 7.3

1   BORROWER(BORROWER-NO, B-NAME, B-ADDRESS, ...)
     TITLE(ISBN-NO, AUTHORS, TITLE, PUBLISHER, YEAR, ...)
     BOOK(ISBN-NO, COPY-NO, ...)
     SUPPLIER(SUPPLIER-NO, S-NAME, S-ADDRESS, ...)
     PURCHASE ORDER(P-ORDER-NO, P-DATE, SUPPLIER-NO, ISBN-NO, QTY-ORD,
       QTY-RECD, DATE-RECD, ...)
     RESERVATION(BORROWER-NO, ISBN-NO, RES-DATE, ...)
     LOAN(BORROWER-NO, ISBN-NO, COPY-NO, RETURN-DATE, AMT-OF-FINE, ...)
     TITLE-BOOK(ISBN-NO, COPY-NO)
Obviously, this was a rather fuzzy question and many interpretations are possible. In accordance with convention, the primary keys have been under-lined in this answer; see section 7.4. Any unique identification of a book

of a title would do; international standard book number is assumed to be convenient. The TITLE entity corresponds to the idea of an item in the catalogue, as distinct from the BOOK itself, which may exist in more than one copy.

2 Yes. In the answer pointer to question 1, there is a relationship with no other attributes. The attributes on LOAN above could be eliminated by attaching RETURN-DATE and AMT-OF-FINE to BOOK; whether one would want to do this may depend on whether one wants to retain a history of past loans made or whether one is satisfied to know only of the current loan.
It is also possible, though less common, for an entity to have no attributes other than its key attribute(s).

3 The definitions all tend to overlap somewhat, except that an attribute must have a value.
It is always possible to substitute an entity-type for a relationship-type, giving the entity-type the attributes that were previously enjoyed by the relationship-type and creating new relationship-types (without non-key attributes) between this entity-type and the original entity-types. This is likely to make a simple relationship, between two entity-types, more complicated than needed for easy comprehension, but may make a complicated relationship, for example involving more than two entities, simpler to understand.
It is always possible to take out an attribute and make it into an entity-type (related to the original entity-type if the attribute was drawn from an entity-type). This may make the model more complicated or, possibly, simpler.
It is always possible to substitute a relationship-type for an entity-type by making all the attributes into entity-types and joining them with the relationship-type. This may make the model more complicated or, possibly, simpler.
What is defined as an entity, relationship or attribute is a matter of usage or custom. The entities etc. are meant to be an aid to thought about the real world, not a minute and perfect description of it; it is a matter of what is meaningful to the analyst and the users. Given a choice of representations which are equally well understood, equally widely recognised and equally in conformity with the facts, the simpler one is preferable, of course.
The references to entities in the next sections should be deemed to include relationships where the context allows. I have deliberately left the words 'entity' and 'relationship' undefined, so that they can stand either for an occurrence or a type; I believe this will make the passages more readable and that the reader will be able to detect the sense from the context.

4 What formal messages do you have in mind? Please enumerate the alternatives or give me rules for generating the alternatives.

5 Unlikely, perhaps, but maybe so if products did not come in pack sizes specific to each product but rather the pack size could be tailored to the requirements of the customer. If **both** possibilities applied, i.e. customers could specify a pack size but, if they did not, then a standard pack size would apply, then we are speaking of two separate attributes: REQUESTED-PACK-SIZE, which is an attribute of ORDER, and STANDARD-PACK-SIZE, which is an attribute of PRODUCT.

Section 7.4

1 See Figure 7.12 overleaf. If a given product can appear only once on an order, PRODUCT # could replace LINE # in the primary key of ORDER-LINE.

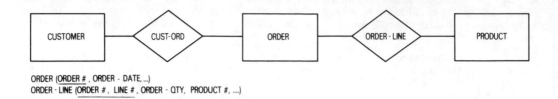

ORDER (ORDER # , ORDER - DATE, ...)
ORDER - LINE (ORDER # , LINE # , ORDER - QTY, PRODUCT #, ...)

Fig. 7.12: Answer pointer to question 1, section 7.4

2   Mine were, as far as can be judged from the named attributes and declared assumptions, and assuming that AUTHORS is regarded as a single string and not an attribute whose value occurs with each author of a multi-author book. Had that been the case, I would have been inclined to call the attribute AUTHOR.

Section 7.5

1   9   A course may not exist without a regiment;
    10   many courses may be prepared by a particular regiment;
    11   a regiment may exist without a course;
    12   more than one regiment may prepare a particular course;
    13   a regiment may not exist without a soldier in it;
    14   only one regiment may hold a particular soldier;
    15   a soldier cannot exist without a regiment;
    16   many soldiers may belong to a particular regiment.

2   Entities and relationships: Created by; Deleted by.
Attributes: Recorded by; Modified by; Authority to inspect.

REFERENCES

(1)   Date, C. J., **An introduction to database systems**, Addison–Wesley, 1975.
(2)   Deen, S. M., **Fundamentals of data base systems**, Macmillan, 1977.
(3)   Chen, P. P., The entity–relationship model – towards a unified view of data, **ACM Transactions on Database Systems**, 1, 1, 9–36, March 1976.
(4)   British Computer Society, **Data analysis for information systems design**, Proceedings of a conference at Loughborough University, June 1978.
(5)   Couger, J. D., Evolution of business systems analysis techniques, **Computing Surveys**, **5**, 3, 167–98, September 1973.

8 Business procedures design

8.1 MODES OF WORKING

The design of business procedures for data processing (DP) systems involves defining the processes which humans are going to apply to the data they deal with, either prior to it entering the computer or after it leaves the machine as an output. Naturally, reponsibilities for defining the procedures should reflect organisation policy on paticipation by employees in the design of work systems. The definition should cover:

a) procedures by which user employees may discharge their responsibilities for creation/deletion/modification/inspection of records concerning entities and relationships in the business;
b) actions to be taken by users in response to reports or other messages emanating from the system.

The design will be constrained by the data model (for example, continuing from the previous chapter, if every soldier must have a number and a name, the procedures should ensure that a name cannot be recorded without a number, and vice versa). Of course, there may also be other constraints or influences on the design. The organisation of the user departments and the design of user procedures is taken in this chapter to be a responsibility of user management and the users themselves. The topics treated here aim to help the analyst with some of the decisions in which he is likely to participate and with specific features of computer input and output design.

One of the basic decisions about procedures is whether the DP operations involved with an event that has happened, such as the placing of an order or the recruitment of a new soldier, will be dealt with there and then – 'transaction mode' – or whether data about similar transactions will be accumulated for processing in one go – 'batch mode'. The transaction mode of working usually (but not necessarily) entails interactive, online working between men and the computer system; direct data input (by keying, or less common means – optical scanning of typed or handwritten characters, voice keywords, etc.); immediate feedback on the acceptability or otherwise of the transaction; performance of all the actions associated with the transaction (i.e. updating all the records affected before turning to the next transaction). The batch mode usually (but not necessarily) involves offline working; indirect data input by keypunched cards, key-to-tape or key-to-disk, optical reading etc.; delayed feedback on acceptability; subjecting the whole batch to successive operations (e.g updating one type of record by all transactions before turning to the next type of record); division of labour.

Question

1   Contrast batch mode work with transaction mode work on the following criteria:

a) currency (age) of data in the database;
b) suitability for a mobile worker (e.g. travelling salesman);
c) cost of hardware for capturing data by keying;

d) likely effect of breakdown of computer;
e) need for redundancy in computer power;
f) need for standby hardware;
g) amount of online storage capacity needed by the computer.   (10 min)

8.2 JOB SATISFACTION

Five contracts view   Mumford (1, after Talcott Parsons) has analysed the idea of job satisfaction by regarding the relationships between employer and employee as a series of contracts. These contracts are not necessarily enforceable contracts in the legal sense, but more informal, tacit agreements about the exchanges between employer and employee. The contracts are as follows.

Knowledge:	employee lends his knowhow to employer; employer trains and develops employee.
Psychological:	employee invests his interest in meeting the ends of the organisation; employer gives the employee psychological rewards, e.g. sense of achievement, recognition, responsibility, status.
Efficiency (rewards):	employee delivers quantity and quality of labour, accepts control; employer gives money rewards, suitable working conditions, task information and supervision.
Ethical (social value):	employee goes along with the values and ideology of the employer; employer's behaviour conforms to that thought ethically acceptable or socially desirable by the employee.
Task (structure and variety):	employee accepts constraints on his job caused by the nature of the work; employer provides variety and interest in the tasks assigned.

  The job satisfaction of an employee is governed by the 'bargains' he has made across-the-board on these contracts. Satisfaction is measured by the fit between the actual state of the set of bargains, and the states expected, aspired to or needed by the individual employee. Since an individual's expectations, aspirations or needs may change over time, it follows that his job satisfaction may also change, even though the bargains remain the same.

Long-term/short-term view   If a job is designed to enlarge the present satisfaction of a particular incumbent, then a short-term view is being taken, since in the long term the particular incumbent, or his expectations etc., may change. Taking the long-term view, i.e. trying to maximise satisfaction of future incumbents, catering for future expectations etc., is somewhat problematic and may contribute to less satisfaction in the short term. The moral is that it is desirable to design jobs as flexibly as possible so that opportunities are left to adjust the balance of a bargain on a particular contract. This is obviously not easy in the case of the Ethical contract. For some features of the other contracts, changes can be made as and when required to meet new circumstances (rate of pay, for example). However, such changes to many other features (task variety, for example) are not easily made unless flexibility has been planned and built into the work system.

Analyst's influence   The systems analyst's ability to influence the contracts is constrained. His terms of reference may require him to deal only with

DP systems and their repercussions within the organisation. Other individuals such as user managers exert more power over the remaining areas of organisational activity.

If a job exists only to serve the DP system (i.e. the DP-related content of the job is 100%), such as in the case of an order entry clerk who encodes written orders for transmission to a computer, then clearly the analyst is in a strong position to influence the contract bargains (although it must be said that in practice many analysts fail to recognise that they have designed the job of another person, or they fail to realise the extent of the responsibility they have assumed by so doing). The analyst can exert his influence directly, through design of the order entry DP system, or indirectly, by making representations to the managers on matters on which he can claim expert status. If the DP-related content of the job is small, as with a night-watchman whose only DP-related task is to clock on and off, the analyst's influence extends no further than to a tiny portion of the task contract.

Even when a job is high in DP-related content, there may be structural constraints on the analyst's influence. The analyst's terms of reference usually drive him through one 'application' after another. With some jobs, each application may contribute to only a slice of the DP-related portion of the job. The whole DP-related portion is made up of slices contributed by different applications developed by different analysts at different times. Consider, for a thick-slice example, a stock replenishment clerk whose job entails monitoring stock levels and re-ordering stocks from suppliers. The monitoring side of his job may be altered at one time by the Stock Control system design team, and the re-ordering side may be altered at another time by the Purchase Orders system design team. A thin-slice example is the job of a data control clerk whose duties are to check and take action on control information for all applications.

Functional job analysis   Another perspective of analyst influence on the Task contract can be got from Fine's proposal for classifying job functions on three dimensions (2): Dealing with People, Dealing with Things and Dealing with Data. Apart from proposing the classification of jobs according to the time spent in each type of dealing, Fine offers a taxonomy for classifying the skill levels that jobs demand on each dimension, running from, at the lowest level, Taking Instructions/Helping People, Tending Things and Comparing Data to, at the highest level, Mentoring People, Precision-working Things and Synthesising Data. The complete taxonomy for Dealing with Data is: Comparing; Copying; Computing; Compiling; Analysing; Coordinating; Synthesising. At a level below Comparing, people are only Observing or Learning about dealings with data.

The analyst's main sphere of influence is in users' Dealings with Data and, in particular, that subset of Dealing with Data which comprises Dealing with Formal Messages. If the job is not predominantly Dealing with Formal Messages, the analyst is comparatively powerless to enrich it (or, for that matter, to impoverish it). This serves to emphasise that the responsibility for job satisfaction must largely rest elsewhere, maybe with the manager who recruits for the job, maybe with members of a staff function like personnel management.

A broad understanding of the profile of a job in terms of skill level and time spent in each dimension is useful to the analyst when exploring reactions of job incumbents to possible system design alternatives.

Aggregate job satisfaction   Although it has just been argued that there are no absolutes in job satisfaction, it is probably a fair observation that rather large numbers of the population in the indutrialised West have jobs which fail to meet their expectations, aspirations and needs. Two factors at work may be surmised, speaking in the aggregate; firstly, rationalisation of work systems (to win efficiency gains through division of labour and production-line processes) has deskilled many jobs to the level of the lowest common denominator in the labour pools, and secondly, increased education and communication have tended to raise expectations and aspirations.

This aspect is confused by the possibility that many workers have adjusted their expectations – downwards – to match their awareness of current job possibilities. Sociologists speak of the 'self-investment' of a worker in his work. A worker who invests much of his 'self' in his work is one who judges himself on work-related factors such as promotion, proficiency, esteem of colleagues, achievement, output. A worker who invests little of his 'self' in his work is one who has his centres of interest outside his employment and who does not judge himself on promotion, proficiency, etc. Job satisfaction is likely to be sensitive to self-investment.

There is also a political dimension to this discussion which may perhaps be revealed by the following two viewpoints.

Viewpoint A   Some people have more satisfying jobs than others. No one is ever completely satisfied, nor should he be. It is degrading to deny a person the opportunity to better himself by his own efforts. People who are not at a disadvantage should be expected to help themselves by competing for the more satisfying jobs. People who are not prepared to invest themselves in this way should not get on. This competition benefits society by selection of the most able and hardworking people to do the most responsible jobs, giving more production and leaving more surplus to distribute to the disadvantaged. Technological benefits should not be thrown away before it is found out whether or not people can adapt to new technology. Viewpoint B drags everyone down to the lowest common denominator and wraps them in the protective cocoon of an engineered society, within which they will stagnate.

Viewpoint B   Some people have more satisfying jobs than others. People have intrinsically equal rights to job satisfaction. It is degrading that intelligent, honest people should apply themselves like cogs in a machine to routine work. Jobs should be designed according to social criteria that assure humanised work, then technology should be adapted to support those jobs. Even if this means sacrificing technical efficiency (which is not conceded), Western society is wealthy enough to stand the loss. People will invest more of themselves if their work is humane. Viewpoint A leads to the formation of a privileged elite intent on defending their positions.

Of course, if a person believes one of these sentiments, it does not follow that he shares the whole viewpoint.

Questions

1   Should a systems analyst decide whether jobs lack satisfaction?   (10 min)

2   If a job lacks satisfaction, should an analyst act on his own initiative to improve it?   (5 min)

3   Why should an analyst 'explore reactions of job incumbents to possible system design alternatives'? Or should he not?   (3 min)

8.3 ENRICHING AN IMPOVERISHED JOB

Should it be judged that a job is impoverished, then the following are enrichment possibilities.

1   Make the job more meaningful to the jobholder. Let him identify how 'his' work is related to the outside world. Instead of having clerks deal with all types of work equally with other clerks let each clerk have a geographic area, a type of client or type of business, even a letter of the alphabet, etc. that he can consider his own. Avoid so much division of labour that the worker can no longer relate his effort to the useful end-product.

2   Give the jobholder a customer. Let the jobholder have direct contact (not through an intermediary such as a supervisor) with the person who uses his end-product, so the jobholder 'accepts the orders' and 'delivers the goods'. The 'customer' need not be an outsider – he may be in another department or section.

3   Draw in ancillary tasks. Look at the tasks performed by others before and after the jobholder's tasks. Include them in the job if they would add interest, increase variety or increase meaningfulness.

4   Draw down control tasks. Look at the tasks performed by the supervisor. Look for ways of delegating increased responsibility and authority to the jobholder.

5   Draw out undesirable tasks. Look at the tasks performed in the job. See if the disliked or meaningless tasks can be delegated, mechanised, dropped. (But do not equate disliked/meaningless with 'routine'. Routine tasks can be liked and meaningful.)

6   Give the jobholder direct feedback. Arrange things so that he knows the results of his own work (or failing that, his own group's work) so he can feel a sense of achievement for good performance and a sense of failure for poor performance.

7   Give the jobholder hope. Look for possibilities of promotion, increased responsibility, special privileges for jobholders with high achievements.

The following example shows these ideas in action. The keypuncher's job is not entirely typical of business-procedure jobs, being very thin-slice usually, 100% DP-related and (especially when keypunchers are employed in the data processing department) comparatively easily influenced by systems analysts. However, the thin-slice problem is usually the most intractable and thin-slice jobs are often quite impoverished, so this case is an especially interesting one.

Before enrichment   Fifty keypunchers operated in a pool. There were ten sections, each with a supervisor. Arbitrary batches of data to be punched, each about an hour's work, were passed from Data Control section (who had checked for legibility and conformity with standards) to the ten supervisors, in equal proportions. Data Control marked the batches with priority. Each supervisor gave the next-highest-priority batch of punching to the first free operator, ensuring that each operator always had work to do. The punched batches were then submitted to another operator within the section, for verification. Errors detected during verification were passed to the supervisor, who allocated them for repunching to the next available operator. The operators had strict instructions to 'punch what they see' even if they thought it was wrong. If a substantial query arose, the operator would return the batch to the supervisor, who would sort out the query with Data Control or the user department. Completed work was returned to Data Control who submitted it to the computer and sent the results to the user department.

After enrichment   There were still ten sections, each with a supervisor, but each operator, or pair of operators, was given continuing responsibility for all the data coming from a particular section in the user departments. The sections were initially allocated on the basis of average workload. The operators themselves checked for legibility and standards, and contacted their customer sections directly with any problems. The operators set their own work schedules, were given some flexibility in when they could start and finish their shifts, and agreed priorities directly with their customers. Work was verified by the punching operator or her partner, and the punching operator always punched her own corrections. Operators were given authority to correct obvious coding errors. Operators submitted their customer section's work to the computer, received error reports, arranged the correction runs and delivered the results.

The computer kept a log of key depressions and errors, and the log was reported to the operators. Any operator who attained a low enough error rate was thereafter allowed to submit her work directly to the computer, without verification. If an operator was overloaded with work, it was the supervisor's job to find an underloaded operator to give temporary assistance, or to arrange a crash team effort on a rush job. Senior and proficient operators were involved in the selection, orientation and training of new staff, designing of operator training programmes, making hardware surveys, trials and recommendations. Some progressed to new jobs in their customer sections.

Many jobs have been made more routine as a result of rationalisation by systems analysts who have overlooked human needs for variety, challenge and responsibility. This is especially true of clerical workers – in a survey by Eason, Damodaran and Stewart (3), 61% of clerks reported increased routine as a result of computerisation.

Questions

1   What effect do you think a job enrichment programme such as that described above would have on keypunching productivity?   (5 min)

2   Contrast batch mode work with transaction mode work on the following criteria:
a)   feelings of involvement with end product;
b)   variety of work for an employee;
c)   challenge or difficulty of work for the employee;
d)   satisfaction of employee's feedback needs.   (5 min)

3   Can you make an alternative proposal for the job enrichment case described above?   (5 min)

8.4 TYPES OF USER

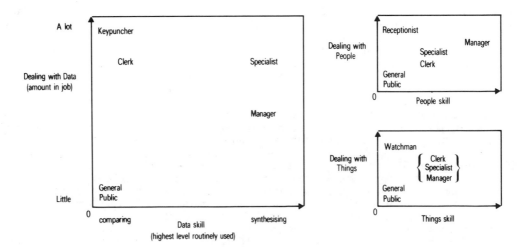

Fig. 8.1: Types of user. Content against skill used in job

The main diagram in Figure 8.1 has Fine's levels of Dealing with Data on the horizontal axis and the amount of Dealing with Data on the vertical axis. Some job-types are positioned on the diagram; for the purpose of this discussion, their positions may be taken as **defining** the names. The inserts show possible positions for these names on the other dimensions, in their ordinary usage, but only the data dimension is of present interest. 'Specialists' comprise quite a large number of specific job-types, including engineers, management scientists, economists. General Public is shorthand for a class of users who, it must be assumed, have little experience or skill and against whom systems must be 'foolproof'.

Figure 8.2 gives examples of how users may have more or less discretion in their direct dealings with the system. By and large, increasing discretion

INPUT	Predefined format and content	Alternative formats	Alternative formats and content	Free format	Query language	Programming language
OUTPUT	Predefined reports at programmed intervals	Predefined reports on request	Parameterised reports		Ad hoc information retrieval	Ad hoc reports

No discretion	Some discretion			Much discretion	

Fig. 8.2: Meaning of discretion over input and output

means increased flexibility (where flexibility is how easily a system can adapt to new requirements or to the idiosyncrasies of a new user). However, discretion generally also brings complexity. More freedom of choice over input and output messages in a **formal** message system means spelling out more alternatives and/or giving more rules prescribing possible actions.

It is probable that users for whom Dealing with Data is a large part of their job (especially if Dealing with Data is taken to mean Dealing with the DP system) are more willing to invest their time in learning all the alternatives and rules, assuming that this results in more satisfaction from the system. Satisfaction means here how well, as perceived by the user, the system's facilities meet the user's requirements and preferences. Users who do not deal much with data or the DP system may regard each extra choice as a nuisance and each extra rule an obstacle to easy use.

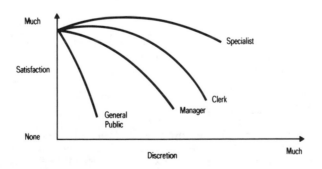

Fig. 8.3: Hypothesised effect of (potentially useful) discretion over input and output on different types of user dealing directly with the DP system

In Figure 8.3, it is supposed that Clerks and Specialists will find a system which gives them some discretion increases their satisfaction, whereas Managers and the General Public prefer things to be simpler. It is probable that the Specialists can deal more easily with DP system complexity than can Clerks, since the former spend more of their time exercising high-level data skills than do the latter. Clerk satisfaction is expected to drop with 'much discretion' but Specialist satisfaction remains comparatively unabated or possibly continues to increase.

Figure 8.3 assumes that the discretion allowed over input and output matches some variability in the user's task, or the discretion he has (or wants) over how to fulfil his task. Discretion over input and output which goes beyond this is useless. Managers often have varied and unpredictable needs for information and much discretion over their work. They may need a system which is flexible enough to meet these demands, but they do not want the complexity that comes with much discretion over input and output.

If it is desired to retain flexibility but to reduce the complexity of discretion, two possible strategies are the **default option** and **indirect dealing**. The default option is a version of the system that runs with little or no discretion if the user takes no special action, e.g. predefined reports at programmed intervals. This should satisfy the minimum requirements of the user who does not want discretion. Such a user need not learn about, nor even be told about, the other options. Users who wish to exercise more discretion can, for example, request additional reports or specialise their view of the system by suppressing their default reports, changing their report interval, etc.

Indirect dealing gives the user an intermediary who deals with the DP system. The user gives his requirements in natural language to an expert, situated either locally at the point of need (in the user's department) or in the DP department. The expert exercises the discretion to meet the user's requirements.

A good design principle, which helps to minimise complexity, is to maintain a consistent and uniform style for all applications throughout the organis- ation, including the method of exercising discretion. If the user has to learn different methods of exercising the same type of discretion over different inputs or outputs, he will no doubt complain.

It is also helpful if the user can be given a clear, simple, mental model of how the system works, ideally in his own terms and using concepts with which he is familiar. Preserving the user's mental model sometimes makes the computer system more complicated than may otherwise be the case, but this is usually a small sacrifice for a large gain. People who are unaccustomed to mental manipulation of new concepts, particularly, need an explanation in terms of familiar, everyday things.

Questions

1   Should an expert intermediary be a Specialist or a Clerk?   (3 min)

2   Should an expert intermediary be a person supervised by the Data Processing department or by the user department?   (3 min)

8.5 INTERACTIVE DIALOGUE DESIGN

An interactive dialogue is an exchange of formal messages between a user and a computer system. This often takes place over a visual display unit equipped with a keyboard. The user exchanges messages, replying to prompts from the computer or volunteering messages, over a period of time. At any one instant of time, he sees in front of him a static display. Considerations of the format of static displays are dealt with in chapter 9; here we are concerned with the dynamic aspect.

The user's contribution to the progress of the dialogue may be simply supplying a yes/no answer, or choosing an answer from a prompting 'menu' of possible answers, or supplying an unlisted answer in reply to a prompt, or supplying an unprompted message. Unprompted messages include supplying a data value together with its name (e.g. EMPNAME=SMITH, PAYNO=147243), or unnamed values in a predefined sequence, or supplying commands for action, either in a fixed format or in a freer command language following certain syntax rules. The last case, which concerns the design of programming languages, is outside the present discussion.

The prompts may take the form of a natural language question, a coded question or a prompt where the user is given a static display of names and empty fields and is expected to fill the values in the empty fields, in a similar way to filling in a form. Commercial displays often allow the 'name' portions of the screen to be protected so that they cannot be corrupted by the users; a similar effect can be achieved on an unbuffered or intelligent terminal by having programs fail to display ('echo') input characters when they are entered at the wrong point, automatically moving the cursor to the correct point. A more explanatory prompt is needed when the system is used infrequently or the users are untrained.

It is sensible to allow all variants of yes and no answers, i.e. YES, Yes, yes, NO, No, no, Y, y, N, n. As a general rule case alternatives should be allowed in any reply where the case shifts are not significant.

When a menu is presented, the user in effect wants to point to an item on the list. He can do this directly with a light pen or with menu selector buttons; a similar effect can be achieved with voice keywords if voice input is available. Failing these, numbering or lettering the items will permit the user to key only the selection number or letter, rather than having to key the words in the menu. Large menus can be split by asking preliminary yes/no questions or they can be removed from the dialogue and placed in a hard copy reference list available to the user (item numbers can still be used on this list, of course). Experienced users could be allowed to shorten prompting messages by specifying they should appear in abbreviated form until further notice, or to by-pass the prompts by entering a string of replies separated by some delimiter character such as an oblique stroke or colon (choose a character in the same case shift as the messages). The user may be allowed to further customise the procedure to suit his preferences by, for example, being able to switch off audible tone signals or flashing fields. Less experienced users could be allowed to call for more helpful and explanatory prompts than usual. The abbreviating, by-passing, customising and helping routines should be standard throughout the organisation.

Dialogues can be seen as a tree which forks at the yes/no points and which has multiple branches at those menus which lead to alternative continuations (as opposed to menus which continue to the same next prompt irrespective of the answer made — such menus are simply gathering data and not changing the course of the dialogue); see Figure 8.4.

It is often sensible to assume that the user will want to make several successive transactions of the same type, in which case the last prompt (e.g. in Figure 8.4, 'Enter credit (small rebate format)') can be reiterated when the entry is concluded. There should be a simple way for the user to abort his entry, even when he is half-way through, and start again. If the user sends no details (e.g. he transmits an empty line), the computer can 'back up' the tree to the preceding prompt. With large trees, it may be desirable to give the user a standard option for getting back to the top of the tree, or even to make direct leaps to other parts of the tree.

Other hints for dialogue design (after Gaines and Facey, 5) follow.

> Give quick response to each input. This may simply be the next prompt, but if time is to be spent calculating, retrieving, etc. a standard symbol should be output indicating that action is being taken on the response; or the type of action being taken should be confirmed. If delay is possible, users will be happier if they are not left 'hanging', not knowing what is going on; confirmation of continued action should be given periodically if possible, or users should have the chance to interrogate the computer about the status of a task given to it.

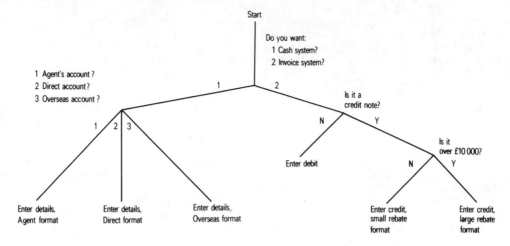

Fig. 8.4: Example dialogue tree

The terminology as well as discretionary choices should be uniform across applications.

Do not have messages in the system which are not clearly meant for the user, for example, operating system messages not requiring user action should be suppressed or routed to operators.

Do not make the user rely upon his memory, e.g. print out the current value of a variable if inviting the user to enter an amended value.

Do not forget that hard copy might be more useful than a display.

Keep records of system and user responses for later analysis when considering system improvements.

Give as much thought to error messages and error corrections as to the main dialogue. A particularly powerful, but simple, error 'message' is to decline to echo an invalid character (e.g. alphabetic character entered in a numeric field), retaining the cursor at the data entry point. This gives the impression of being 'unable to get the data in', rather like being unable to get a coin accepted by a slot machine.

Do not assume the user gives his full attention all the time. He may be distracted, or leave his station, at any moment. No fleeting messages.

If a user sends a non-printing character (e.g. function button, cursor move), remember that it may be meaningless to send an error message about an invisible character.

Questions

1   What records would you keep of system and user responses?   (3 min)

2   Suppose a user has to submit details of customer-name, customer-number, credit-requested, date. What form would the dialogue take with (a) a prompting question style, (b) an unnamed list, (c) a named list, (d) a coded prompt, and (e) form-filling?   (10 min)

3   Some of the types of dialogue could be summarised as follows:
1   natural language prompt;
2   coded prompt;
3   yes/no answer;
4   value answer;
5   name and value message;
6   unnamed value message;
7   command keys;

8   command keywords;
9   command language;
10   menu selection by button;
11   menu selection by selection number/letter;
12   form-filling;
13   display update (the operator is given an entire record on the screen, he updates some portion of it and sends back the whole record).

These types by no means exhaust the possibilities and are not mutually exclusive (for example, form-filling generally consists of natural-language prompts followed by value answers, and may include coded prompts, yes/no answers, keywords, menu selection).

Which types would you pick as best suited to:
a)   A General Public user?
b)   An inexperienced keyer?
c)   Speedy progress of the dialogue?
d)   Large volume data, input by experienced operators?

Which types would you pick as least suitable for:
e)   A slow terminal?
f)   A Manager?   (10 min)

4   How would you document your design of:
a)   A question and answer data-gathering dialogue (no branches)?
b)   A question and answer dialogue with branches?
c)   A form-filling display?
d)   A question and answer dialogue with branches to alternative form-filling displays?   (10 min)

5   Despite all your efforts, the users may think your dialogue too complex or too simple or too cumbersome. You may accidentally slip into a computer jargon use of a word in a prompt and so be misunderstood. Some users may like your dialogue, others not. Some may change their minds when they get to know it better. What precautions should you take?   (3 min)

ASSIGNMENT

Define an algorithm which will accept a date from a user in any form he chooses, assuming it is one usually acceptable to humans, and convert this to a standard internal format YYMMDD.

The English usually present a numeric date in the form DDMMYY, but the Americans favour MMDDYY. If your algorithm has to cater for both Americans and Englishmen, what complications does this bring?

This seems to be a case where the discretion possibilities do not have to be spelled out to users since hopefully all natural language variations are catered for. Are there any disadvantages?

ANSWER POINTERS

Section 8.1

1 a) Batch older.
b) Batch more practical, but maybe the travelling salesman would prefer to conclude all his business at each sale, transaction fashion, if this were technically feasible.
c) Transaction mode has cheaper purchase price per station (but more stations or other hardware/software enhancements may be needed).
d) Batch mode decouples computer operations from business operations, so little effect; transaction mode tightly couples the two, so business operations stop unless some standby system is available.
e) Transaction mode needs larger computer to handle peak loads of work with acceptable response time; batch load can be more evenly spread.

f)   Transaction mode has more need for standby hardware (see d).
g)   Transaction mode needs more online storage capacity.

Section 8.2

1   The analyst, as a responsible member of society, ought to be concerned about whether the jobs his work is affecting are satisfying or not. But only the jobholder can 'decide', so if there is a need to know, it is the user who should be asked. Although the jobholder may say if it is satisfying or not, he may not be able to pinpoint the satisfactory/unsatisfactory features. People do not always recognise the cause of their dissatisfaction, sometimes blaming a secondary feature. Education about the satisfying and dissatisfying aspects of jobs can help overcome this.
  Although the analyst ought to be concerned, so ought user management. The user management's concern is not only that of the citizen; they also have the duty that any manager has to be concerned about the welfare of staff.

2   The user management have the main responsibility to act (see 1 above) and, if they do, this question may not arise. The analyst is in a moral dilemma if he believes that action is necessary but user management is not taking it. There is no clear-cut right course here. A reasonable middle road would be to record one's opinion and promote open discussion of the issue.

3   This is really the same point as above. The users rather than the analyst should be exploring reactions. The sponsor wants reactions explored so that he can be satisfied that the new system will work as intended.

Section 8.3

1   Davis and Cherns (6) have reports of two measured cases, both similar to this one, which increased productivity - quantity and quality improved, absenteeism reduced.
  The log of key depressions is a matter for concern, since feelings about this may hang on a knife-edge. Presumably the keypunchers saw the log as something which worked to their advantage. However, such a log could just as easily cause discontent by promoting the feeling that Big Brother Is Watching You, especially if it were received by an inconsiderate management.

2   Transaction mode, at face value, is likely to increase them all. However, caution must be exercised with any prediction of human reactions because one is dealing with very complex affairs.

3   Of course, we do not know all the constraints, but possibilities are:
a)   relocate the keypunchers in the user sections;
b)   go to transaction mode, have users (retrained keypunch operators?) enter transactions directly through keyboards/visual displays in user departments.

Section 8.4

1   It depends on the degree of discretion needed over input and output. If use of a query language or programming language, or complex parameters, is called for, a Specialist will be most suitable. If the discretion is to call for alternative predefined reports or to supply simple parameters, a Clerk will be more suitable.

2   Either supervisory method could be made to work. If being an intermediary is a substantial part of the expert's work, there is a good case for placing him in the user department, supervised by user management, since his job exists so that users can be served through a close understanding of their needs. Particularly in the case of Specialist intermediaries, a counter-argument is that technical supervision, training and career development might be better supplied by the Data Processing department.

Section 8.5

1   Numbers of transactions, by type. Frequency with which different choices are made. Frequency of different types of error.

2   (a)

Computer	User
What is customer name?	SMITH
His number?	11726
Amount of credit requested? (£'s)	200
Date of request?	11/2/80

(b)   SMITH;11726;200;11/2/80
(c)   CUSTNAME=SMITH, CUSTNO=11726, REQUESTDATE=11/2/80, AMOUNT=200
(d)

Computer	User
CNAME = ?	SMITH
CNO = ?	11726
AMT = ?	200
DATE = ?	11/2/80

(c)   CUSTOMER NAME |SMITH | CUSTOMER NO |11726|
      CREDIT REQUESTED £|200 | DATE OF REQUEST |11|/|2 |/|80|

3   a) Natural language prompt, yes/no answer, menu selection by button.
b)   Command keys, menu selection by button; yes/no answer, menu selection by selection number/letter.
c)   Command keys; unnamed value message, command keywords, coded prompts. Yes/no answers and menu selections may be quick in themselves but they often make the whole dialogue more pedestrian.
d)   Unnamed value message, form-filling.
e)   Menu selection; display update, natural language prompt.
f)   Command language.

4   a) Two columns: (1) computer displays (2) user replies.
b)   Flowchart.
c)   If there is only one display with a fixed place for error messages, only the display format need be documented (see chapter 9).
d)   A flowchart with cross-references to the display formats.

5   Some possible precautions are: get users to participate in the design; try out alternatives by simulated trials with representative users, or develop a prototype system for evaluation; give users discretion over input by giving them the option of abbreviated prompts or unprompted replies.

REFERENCES

(1)   Mumford, E., **Job satisfaction – a study of computer specialists**, Longman, London, 1972.
(2)   In Dickmann, R. A., **Personnel implications for business data processing**, Wiley, 1971.
(3)   Eason, K. D., Damodaran, L., and Stewart, T. F. M., **A survey of man-computer interaction in commercial applications**, HUSAT report no. LUTERG 144, University of Technology, Loughborough. 270 pp., 54 refs.
(4)   Eason, K. D., Damodaran, L., and Stewart, T. F. M., **The effective use of computer systems by non-expert users**, HUSAT report no. LUTERG 157, University of Technology, Loughborough. 71 pp., 28 refs.
(5)   Gaines, B. R., and Facey, P.V., **Programming interactive dialogues**, in Parkin, A. (Ed), **Computing and People**, Edward Arnold, 1977.
(6)   Davis, L. E. and Cherns, A. B., **The quality of working life**, Free Press, New York, 1975. 2 vols., 837pp.

9 Design of message forms and displays

9.1 A WORKABLE PROCEDURE

The design of input and output forms, reports and other visual displays for man–computer and man–man communication of formal messages is possibly the most creative task the analyst is asked to do. Arguably, it is also the most difficult to do well. It is possible for systems to limp along with with inelegant, sub–optimal forms and displays; it is for the reader to judge whether this is fortunate or not.

The target is to communicate a formal message visually. 'Communicate' is not quite a broad enough word in ordinary use. I must define it to include making a record for possible later retrieval, perhaps by an unknown hand for an ill–defined purpose. There is rich variety in the design possibilities: different media and technical alternatives, input–output device alternatives, different layouts and aesthetic possibilities. There is rich variety, too, in the influences possibly constraining a particular design: the purposes; the conditions under which the message is communicated; the available technology; established practices and house style. The objective of this chapter is to offer an outline workable methodology for design, but with no pretentions to completeness.

Involvement of the users of the messages is particularly important. The following steps are proposed assuming user representatives are participating as partners.

1   Research the purpose.
2   Make a decision on the message medium (e.g. plain listing paper, pre–printed form, visual display unit, microfilm, card etc.).
3   Design the layout (and the make–up of paper documents) including the layout of copies, if any.
4   Check out the design with a local expert, e.g. check out a paper document with a printer; pick his brains. This may result in going back one or two steps.
5   Test a mock–up of the design in operation. This may result in going back two or three steps.

The following sections are principally concerned with steps 1 and 3. For additional advice on paper documents, the booklet by Wiggins Teape is recommended (1).

Question

1   Do the aesthetic qualities of a form or display matter?   (3 min)

9.2 THE PURPOSE

Messages, it may be supposed, have primarily a semantic purpose: the communication of meaning from the sender to the recipient. The originator of a message destined for a computer is concerned with the semantics of the message, e.g. an order clerk taking an order and completing an order form. The end user of a message emanating from a computer is also concerned with the semantic purpose, e.g. a customer receiving a bill. In between, the message may be **processed** by machines and people who have no concern for the semantics of the message, e.g. a data control clerk who logs batches of forms in and out, a secretary who blindly photocopies a document for the manager, a keyboard operator who is told 'type what you see'. Minimisation of this non-semantic processing by humans is a good design aim.

Where a message is passed from an originator to a human intermediary processor, the analyst must consider both the ease with which the originator can record the message and the processor's ease of processing. Where a message is passed from a computer to a human recipient, the ease with which the recipient can abstract the meaning is important. Where a message is intended to promote action on the part of the recipient, the analyst should also be concerned with how effective the message is as a motivator. The designer's target is therefore to design a message - usually, a visual message - having regard to

the semantic purpose,
the motivational effectiveness, and
the ease of processing

for all the **humans** affected by the message. Ease of processing by computer or other machine is not usually a worthwhile target to pursue if it detracts from these other targets; nor is minimisation of forms costs. One-time costs of design, artwork and ordering, and media production costs such as printing, paper, benchwork (collating, gumming, perforating etc.) are generally small compared with the cost of failure to communicate the meaning precisely, failure to promote the desired action or failure to permit easy processing.

To analyse these aspects of message use so that the design aims are clear, the **message cycles** can be investigated, either forwards or backwards, as in Figure 9.1.

The message cycles are origination, processing and reception.

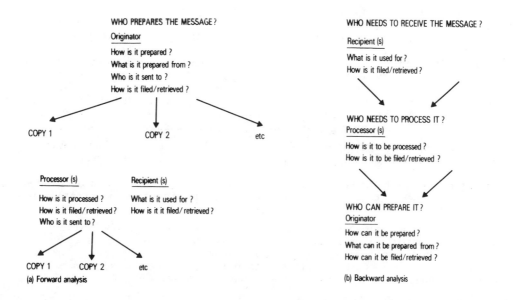

Fig. 9.1: Analysis of message cycles

Origination: originate (record) message;
               transmit and/or file message;
               retrieve filed messages.
Processing: receive messages;
               process message (e.g. transcribe, key, reproduce, handle,
                 absorb);
               transmit and/or file or discard message;
               retrieve filed messages.
Reception: receive message;
               file or discard message;
               retrieve filed messages.

A forward analysis, from origination onwards, is generally most helpful when redesigning messages for an existing system. The backward analysis is more suitable for provoking ideas when a new or redesigned system is involved.

Question

1   In Figure 9.5, what do you imagine are the forward message cycles of the cheque requisition?   (10 min)

9.3 LAYOUT DESIGN

The message will concern the attributes of one or more entities or relationships, so the designer must have to hand details of the attributes which are to appear. The details needed are the type of data which appears for an attribute (e.g. financial amount, date, key, percentage, period, quantity, free text, time), how the data is to be edited and what the maximum field width will be after editing. (This editing information is sensibly recorded in the data dictionary or on the Data Definition form of the attribute in question: see section 6.4.) In addition to the attribute **values** and other values derived from the attributes or incidental to the message, the message may contain constants in the form of a document title, page headings, explanations, other information or instructions, and attribute **names**. The attribute name is not necessary where context, convention, or the value itself make it clear to which attribute a value refers (e.g. customer name and address on a statement). An attribute name may also be omitted when the originator/recipient concerned with that attribute is **trained** (possibly by experience) to recognise the attribute from other cues, e.g. position of the attribute on a layout or sequence in a list. In all other cases, the attribute name will need to be displayed, either to identify an output attribute or to indicate which input attribute is to be recorded.
The obstacles to conveying the meaning of the message could be:

1   bad choice of attribute name to be used to label the attribute values or bad choice of other constants such as titles, page headings, explanations;
2   clutter, i.e. otherwise obscuring the meaning by, for example, distracting attention away from a field of interest or failing to concentrate attention on this field.

Names and other constants should be as short as possible without loss of meaning or quality of presentation. General Public must be assumed to have a more limited vocabulary than Clerk, Specialist or Manager. The analyst should consult Roget's Thesaurus (2) to help him pick the best word.
Some suggestions for reducing clutter are as follows.

1   Consistency across layouts. A standard place for form titles, form numbers, instructions and explanations; a standard house style; standard

relationships between names and values (e.g. values alongside/under names); standard terminology. All these points help to reduce distractions. Related messages, e.g. order form and invoice, should have similar layouts.

2   Eliminate attribute names. Unnecessary names can be omitted. A list format allows several messages (or the 'many' part of a message concerning a one-to-many or many-to-many relationship) to be recorded in columnar fashion, with attribute values under column headings containing the attribute names once only.

3   Distinguish attribute names from attribute values. This can be done, for example, by using lists, printing names in a different typeface, using coloured or shaded backgrounds or reverse video for names, tabulating name-value pairs so that the names are aligned vertically on one tabulation and the values aligned on another.

4   Juxtapose related name-value pairs. Related items (e.g. delivery address, delivery instructions, carrier's name) can be grouped horizontally, vertically or in blocks.

5   Place key fields and totalling fields strategically. Key fields convention-ally appear in the top right-hand corner of documents or at the beginning of a line when a list format is adopted. A key field used for manual filing should be easily locatable when the document is filed. Totals are usually placed below the values they total and totalled values are usually placed to the right of a layout.

6   Avoid proliferation of typefaces and sizes. Two faces, one for names and one for values, are usually adequate. The name style may be used in two or three sizes or with a bold face to allow for headings, etc.

7   Use case thoughtfully. Injudicious use of upper and lower case can bring its own particular form of clutter. A fair general rule is to use lower case for narratives where possible; lower case for attribute names may also be marginally more pleasing or effective than upper case (use a capital letter at the beginning of a word or phrase in accordance with the usual rules of grammar). For example, an attribute name in lower case – Tax to date – may be preferred to TAX TO DATE or Tax To Date. Where a reader is searching free text for a particular keyword, though, there is evidence that a faster search is made if the whole text is in upper case (presumably because this reduces the variety in the target). Also, if attribute names are generally short and names are distinguished from values by other means such as alignment, upper case names may be marginally preferable. Of course, many input-output devices are restricted to upper case, and in many established systems there is no case information in the data held by the system.

8   Use good abbreviations. Abbreviations should be avoided when General Public is concerned, but Clerk, Specialist and Manager will all tolerate abbreviations or even find them advantageous. With regular use and familiarity, even a cryptic single-digit code can be used to convey as much information as a lengthy narrative, greatly speeding recording, transcription and reception. Abbreviations should be acronyms or mnemonics rather than cryptic codes, should follow some consistent rule if there are a number of related abbreviations, and should follow or be consistent with established custom where abbreviations are already in use. See chapter 10 for further discussion of codes.

 Unfamiliar abbreviation of attribute names should be avoided, although this can sometimes be difficult. For example, a column-headed list may concern an attribute whose name is long but whose value is short. The column width should ideally reflect the value length, not the name length. With pre-printed forms, a smaller type size or re-orientation of the headings to a 45° slope can solve this problem. Where only one type size/orientation is available, though, the problem is more severe. Sometimes it is possible to re-orientate the layout so that the 'list' goes from left to right, the headings being listed down the left edge. Failing this, the alternatives are:

abbreviations;
splitting the heading over two or more lines;
using a code symbol with an explanatory key elsewhere on the page.

Running the headings vertically is not usually advisable.
9   Leave margins. Apart from improved appearance, margins may serve to allow data to be read when a document is filed, reduce difficulties caused by torn edges or faulty decollation/bursting of paper, and avoid the distortion at the edges of cathode-ray-tube visual display units.
10   Highlight the features. Key values, actionable values, priorities, related groups of items, totals and other important values, etc. can be highlighted by:

colour; shading; reverse video;
type size; type face; bold face or italics; case - of value or name;
box outlines;
increased tabulation and drop;
flashing field or cursor;
attention-drawing symbols (e.g. asterisks, arrows).

Overkill on these highlights is to be avoided, or they lose their purpose. On fixed spacing-size displays, the thick-space separation of multi-word attribute names may lead to unwanted highlighting of the individual words; hyphenation can reduce this effect.

Displays and forms are planned and documented on a grid such as a print layout chart or display chart (Figure 9.2). The print layout chart grid conforms with standard printer spacings, but the display chart corresponds neither to printer spacing nor necessarily to visual display unit spacing, so allowance must be made for this if considering the finished appearance. A pre-printed form is planned on the print layout grid so that the printer can preserve the correct spacing and drops for the computer-printed parts. For such a usage, the positions at which the computer will print are simply filled with X's. The other use of the charts is to specify layouts for programming; for this use the actual size of the grid boxes is unimportant, the important thing is to preserve the desired number of spaces or lines, etc. so that program output will fit the form or produce the display that was intended.

When documenting a form or display for programming, it is necessary to define all the constants to be produced by the computer, as well as the variables, the editing rules and other features. If the constants are to be 'protected' on a screen (preventing corruption by the user), one convention is to enclose them in a thin-line box which extends around the protected area. The maximum length of variable fields (containing data input by a user or output by the computer) can be illustrated by square brackets on the lines (see Figure 9.2). Specimen data or picture editing characters can be placed in the brackets. Using the latter can obviate the need to define record layouts for each line, but this works only with simple layouts.

Reverse video can be illustrated by lightly cross-hatching the reversed characters. Cursor position and flashing fields are probably best described under 'notes'. Use of colour is not recommended if the layouts are to be photocopied.

Question

1   You have an alphanumeric visual display unit with upper and lower case character sets. It can display 20 lines with 80 characters on each line. You have to display the following message as a menu of attributes and their present values, and enquire of the user whether he wants to have a new display or update one of these values - if the latter, which value. You have reverse video, flashing field and cursor, and screen protect facilities

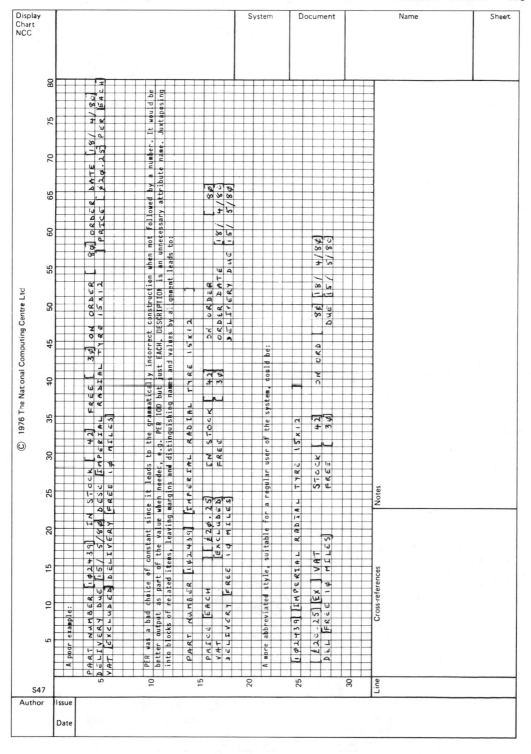

Fig. 9.2: Example visual display layouts on a display chart

available.

   Employee name X(20). Employee address - 4 lines of X(30). Payroll
   number XX9(6)X. Basic pay 999.99. Weekly or mothly paid. Overtime
   pay rate 99.99. Tax code 999A. Paid by cheque/bank credit/cash. NI
   contracted out/in. Special allowances x 4 (allowance code 99, amount
   999.99). Trade union subscription 99.99. Superannuation percent 9.99%.
   Voluntary superannuation amount 999.99. Other deductions x 4 (deduction
   code 99, amount 999.99).

9.4 SPECIAL CONSIDERATIONS - INPUT FORMS

Input layouts (human-originated messages destined for the computer) call
for special attention to recording data, giving instructions and allowing
for continuations.

Recording data   The method of recording should take into account the
circumstances under which the record is made. A message which is to be
originated in a gale force wind on an oil rig in the North Sea imposes
different constraints from one which is to be originated in a serene office.
The neat form pictured by the analyst may be completed on the factory floor
by dirty hands which have nowhere to rest the paper.
   Handwritten data is generally allowed about three lines per inch and
between four and six characters (block capitals) per inch. Clerks may be
able to manage with less space than this; General Public (assuming they
are literate) may need more. Boxes or half-boxes for characters and decimal
points for alignment of numeric fields may be provided if acceptable locally
and a fixed format is desired. Justification and leading zeros should not
otherwise be called for in quantities, dates and currency amounts, although
they may reduce error in codes and other identifiers of fixed length.
Variable-length data may, of course, be embedded in a fixed-length field
when the data is keyed. If this is not desired, then end-of-field symbols
must be keyed. If the sequence of the data items is not fixed, it will also
be necessary to key extra symbols in order to identify the items (this
statement neglects the rather unlikely possibility that the value domains
of the items are all mutually exclusive; if they were, then in theory the
values entered would identify the nature of the item).
   The amount of writing can be reduced by providing menus for ticking or
circling. To reduce keying, the menu selections can be encoded alongside
the tick boxes or on a copy underneath the items to be circled. (Of course,
keying may also be eliminated by optical character or mark reading, or
machine-readable by-products.)
   The values entered for a particular attribute are unlikely to be uniformly
drawn from the value domain. Especially when the number of symbols in
the attribute is large, consideration should be given to providing a menu
for the most common values (e.g. Figure 9.3). Sometimes all the values can
be menued; sometimes it can be worthwhile to enlarge the form size or
provide separate sheets to permit larger value menus.
   Forms to be completed by typewriter need to match typewriter spacing, of
course; generally six lines per inch and 10 (pica) or 12 (elite) characters
per inch in the UK. Typewritten forms should be designed to minimise

Customer account number: ✓

Giant Purchasers S.A.	9	0	7	6	1	A
Big Business Inc.	4	1	3	2	7	D
Global Deals Ltd	7	2	4	7	6	B
Other - enter number ─────▶						

Fig. 9.3: Three very frequent customers

carriage movement and space bar depressions (for example: by arranging menus horizontally with overhead captions, in order of likelihood of being chosen; by aligning fields vertically on tab stops).

Instructions   Instructions (such as advice for completion, distribution, filing, keying, reordering, continuation) intended for different readers should be identified in some uniform fashion (type face, colour, identifying symbol, 'For office use only' space, etc.). Distribution of multi-part documents is aided by colour coding.

The amount and prominence of instruction included should increase as the frequency of completion by an originator decreases. Column identification for keying can be achieved by small or faint print annotations alongside value boxes, by a superimposed mask at the keying station or by a keying copy.

The flow technique (Figure 9.4) can assist completion of forms which contain optional fields which are to be completed conditionally on some other value.

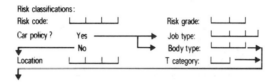

Fig. 9.4: Job type and Body type are required only if car policy is true, location and T category only if car policy is false

Continuation   Where a layout contains a list (e.g. a repeating group such as order lines on an order), provision must be made for the possible exhaustion of the available space (if there is no definite limit on the number of entries). Dividing the message into two messages (e.g. splitting into two orders) is recommended wherever possible. Failing this, continuation must be allowed either on a further copy of the layout or on a special continuation layout. The problems of continuation, such as increased handling, risk of procedural error and need to repeat key data fields from the original layout, are an inconvenience, especially with paper documents. As generous as possible an allowance for repeating groups should be made on the original layout to reduce the need for continuation (obviously, one does not take this advice beyond the point where there is no further significant decrease in the number of continuations needed).

Question

1   It is often a mistake to put instructions on a form, e.g. instructions for keypunch operators to insert a record type prefix before keying the message, where the instructions are cryptic to the originator of the message. Why?   (2 min)

Exercise

Examine the clerical procedure flowchart as shown in Figure 9.5 and make sure you understand it. Design the cheque requisition form in detail, and the combined cheque and payment advice in outline. Use your imagination to supply any missing information you need.

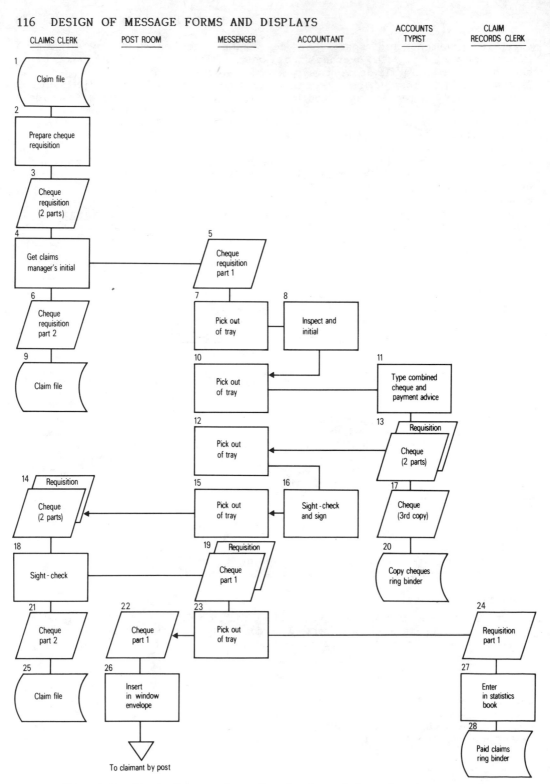

Fig. 9.5: Clerical procedure flowchart – claims cheque processing

9.5 SPECIAL CONSIDERATIONS – OUTPUT

Output layouts (computer-originated messages for humans) also need attention to allowing for continuations, as well as editing, mailing, sequencing and graphic presentation.

Continuations   The continuation of outputs on continuous stationery pre-printed forms does not normally have the option of a special continuation layout. If continued items require special handling, e.g. manual insertion in envelopes when singletons are automatically inserted, it will be desirable to sequence the output so that all continuations appear together (if volumes are at all large). This will also be a reason for minimising the number of continuations.

  The distribution of the number of lines in a variable-length list should be established. The number of lines allowed in the layout can be set to minimise the amount of paper used, if minimisation of the number of continuations is not a target.

  Multi-page output should be page-numbered, and the start and end pages should be clearly recognisable.

Editing   Floating currency signs, check protect asterisks, number of decimal places, decimal point or comma, treatment of zero fields, arithmetic signs, debit/credit symbols •(which symbol for which sense) should be specified.

Mailing and spare space   Naturally, forms to be mailed must be designed to suit the intended mailing method, e.g. continuous postcards, window envelopes, peel-off address labels, transparent wrappers. It is quite often desired to add extra lines of information, slogans or news on material sent to customers. It can be worthwhile to allow blank space on a pre-printed form so that these can be inserted by the computer.

Sequence   Lists should be sequenced so that individual items can be quickly traced and recognised by the recipient. Suppressing the repeated values in the sort key fields of a list reduces clutter by highlighting the key changes that occur.

  The sequence in which successive pages appear off a line printer should also be considered from the point of view of handling and distribution.

Graphic presentation of information   Quantitative information should be considered for graphic presentation, generally as graphs or histograms. Consideration should be given to the problem of overlapping traces on graphs and to providing a grid scale. On colour graphic terminals, a good effect can be achieved with bright colours (yellow, orange, white) for the variable traces, on a blue grid.

Turnaround output   A turnaround output is one which (with or without the addition of more information) is also used directly as input, thus saving keying. Such output is usually in the form of computer-printed documents which can be read by an optical character or optical mark reader.

Questions

1   How would you design the output report illustrated as Figure 4.5? Only a regular line printer is available.   (5 min)

2   Suppose the cost of resequencing and manually inserting continued output exceeds the savings in postage by so doing. Is it reasonable to send continuation sheets in separate envelopes?   (3 min)

ASSIGNMENT

Continuing with the assignment at the end of chapter 7, make a detailed design of the following: an order form for completion by customers of the

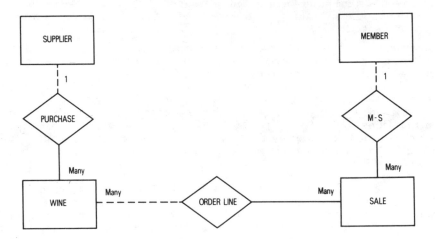

MEMBER (MEMBER #, MEMBER-NAME, MEMBER-ADDRESS, SUBSCRIPTION DUE DATE, CREDIT BALANCE,...)

SALE (SALE #, SALE DATE, SALE/RETURN INDICATOR, RETURN DETAILS (ORIGINAL SALE #, REASON), GOODS AMOUNT, CARRIAGE

   AMOUNT, DISCOUNT (PERCENT, AMOUNT), VAT (PERCENT, AMOUNT), NET AMOUNT, AMOUNT PAID,...)

M - S'(MEMBER #, SALE #)

WINE (WINE #, BIN #, WINE-NAME, WINE DETAILS (YEAR, COLOUR, ...), PREVAILING SUBSTITUTE WINE #, WINE PRICE, WINE QUANTITY

   IN STOCK, BIN END LEVEL, ...)

ORDER LINE (SALE #, WINE #, SUBSTITUTED WINE #, QUANTITY ORDERED, QUANTITY SHIPPED, UNIT PRICE CHARGED, ...)

SUPPLIER (SUPPLIER #, SUPPLIER NAME, SUPPLIER ADDRESS, ...)

PURCHASE (WINE #, SUPPLIER #, PURCHASE DATE, PURCHASE QUANTITY ORDERED, PURCHASE QUANTITY RECEIVED, PURCHASE PRICE PER UNIT ,...)

Feature assumptions:

A wine has only one substitute prevailing at a time; this is changed when the substitute reaches

 bin end .

A wine is supplied by only one supplier.

Membership lapses if subscriptions are not paid by the due date .

After a revised wine list is circulated, only orders on the old list are processed for the period

 of postal delay; then WINE PRICEs are updated and only orders on the new list are accepted.

Parameters needed for messages include: TODAY'S DATE, PREVAILING VAT - RATE , PREVAILING MEMBERSHIP

 SUBSCRIPTION AMOUNT.

Fig. 9.6: Mail—order wine club – possible data model

mail—order wine club; a procedure and layout for keying this into a visual display unit; a document which will allow the warehousemen to pick out the orders; and an order confirmation for despatch to the customer by mail. A despatch note copy of the order confirmation should accompany the consignment. The picking list and order confirmation are to be produced on a regular line printer. An entity–relationship diagram and other information about the system is given in Figure 9.6 should you prefer to use my assumptions rather than the ones you made on the earlier parts of the assignment. You will need to make your own assumptions about the length of codes, etc.

ANSWER POINTERS

Section 9.1

1 For non-Philistines, the answer is an unqualified 'Yes'. For Philistines, the answer is 'Yes – because the effectiveness of the message, how well someone fills in a form, or whether it gets filled in at all, are influenced by the aesthetic appearance'. A tatty, unstructured, roughly typed and duplicated form says everywhere it goes, 'Nobody gave me any thought when I was designed'. Who can blame the users of a form if they give it the same treatment as did the designer?

Section 9.2

1 I imagined:

Cheque requisition

Handled by messenger at each routing

Originator – claims clerk
Prepared by hand from claim details in claim file folder
Initialled by claims manager, who checks it with file details
Copy 2 filed in folder
|
Copy 1 – accountant
Inspected and initialled
|
Copy 1 – accounts typist
Used to provide data for typed cheque and payment advice
|
Copy 1 – accountant
Used for checking accuracy of cheque and payment advice
|
Copy 1 – claims clerk
Used for checking accuracy of cheque and payment advice
Possibly used to identify claim file to be retrieved
|
Copy 1 – claim records clerk
Used for entering claim code, date and amount in statistics book
Filed on paid claims ring binder in date-of-claim order

Section 9.3

1 See Figure 9.7, overleaf.

Section 9.4

1 Increased clutter. The slightly offensive suggestion to the originator that there is a bit of the form that is 'none of his business'.

Section 9.5

1 The variation, variation trend and expected variation are presumably not directly useful but merely intermediate steps towards the seasonally-adjusted sales figure. Similarly, the figure for the last twelve months is presumably redundant for decision-making purposes. It is best to discuss these points with the managers who use the report and explore their reactions to a bar-chart of seasonally-adjusted sales. One possibility would

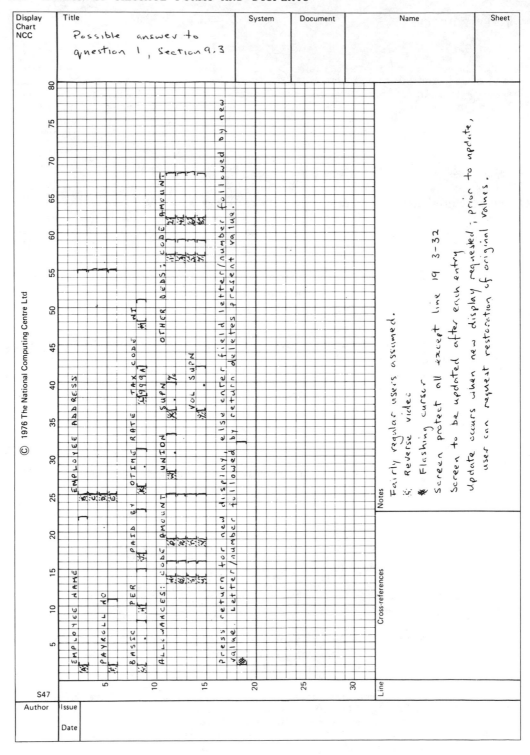

Fig. 9.7: Possible answer to question 1, section 9.3

be to centralise the trend line on the Y–axis (approximately), with the months running along the X–axis, and to show the seasonally–adjusted sales as bars above or below this line.

This would require slightly tricky programming if there were no suitable package or routine already available. Obviously, all this is rather fanciful if the report is not much used, so it should be of concern whether or not the improved presentation is worth the effort. The managers may not realise the effort involved when expressing their preferences, so the analyst has a duty to make the cost clear. Some people go so far as to suggest that report requirements of managers are often fanciful, and suggest the rather drastic remedy of withholding all reports; the complaints that then come will identify the reports genuinely being used.

2   Perfectly reasonable, but (especially in a public service organisation) it might be sensible to explain to the customer concerned why it was done. A use for that spare message space?

REFERENCES

(1) Wiggins Teape Ltd, **A practical guide to business forms,** Idem Division of Wiggins Teape Ltd, 24 pp.
(2) Roget's Thesaurus, 1852. Revised edition, Longmans, 1962.

10 Codes and controls

10.1 CODE DESIGN – HUMAN FACTORS

Codes are introduced into data processing systems for several purposes.

Unique identification   Two occurrences of entities or relationships may have the same set of attribute values recorded in the database. For example, two identical orders may be placed by the same customer on a single day. A code may be assigned to distinguish between such occurrences: order number 1, order number 2. Usually, of course, such codes are extended to provide a unique identifier of occurrences even though the collected attribute values are also unique. So **every** order would be given a unique order number, in the interests of brevity and ease of use as explained below. The coded identifier is usually chosen as the primary key of the entity or relationship relation.

Abbreviation   A coded identifier may be much shorter than the identifying attribute values. Order number is much shorter than customer name/order date/order time. When the attribute can take only a limited set of values, e.g. RED or YELLOW or BLUE, data compression can be achieved through codification, e.g. R = RED, Y = YELLOW, B = BLUE.

Reduced variety, increased uniformity   Humans may exercise discretion over the symbol string of uncoded attribute values, particularly with natural-language items such as names (see section 3.1.1). With made-up codes, it is easier to propose rules which limit this discretion, e.g. the code is of fixed length, only certain symbols may appear. The uniformity that results can lead to increased ease of use or reduced error, in manual systems as well as computerised systems.

Secrecy   An attribute value which is to remain secret may be represented by a cryptic code, i.e. one where there is no obvious relationship between the value of the attribute and the value of the code. Some relationship must exist in order to decipher the code. How well the secret of this relationship is kept, and how difficult it is to discover the relationship not knowing the secret, are measures of the secrecy of the code.

The code, or coded attribute, may be used for sorting, filing, retrieval or matching of records in manual or computer systems. The abbreviation and variety-reduction lead to codes often being used for classifying entities or relationships where such classifications are needed for management reports.

Short of expecting the system to cope with natural-language discretion, the codes in a system should be designed primarily for human ease of use and freedom from error. Ease of processing on the computer is usually of secondary importance.

Ease of use may depend upon preserving, in an abbreviating code, the meaning of the original attribute value. Mnemonic or faceted codes, as described in section 10.2, are preferred for this purpose. If the attribute is solely an identifier, a sequence code is preferable. (If a code is to be keyed by an inexperienced operator, a simple numeric code for keying on a numeric pad may outweigh loss of meaningfulness, by giving rise to less keying error.)

Errors in transcribing or recalling codes may thus be caused by lack of meaning in the code or lack of manipulative skill in keying. The balance of error in transcription or recall can broadly be attributed to limitations of human short-term memory capacity. It is hypothesised, by analogy with computers, that a human memorises codes by first reading the symbols into a register called short-term memory (STM). This has a capacity of about seven symbols. The symbols may remain in short-term memory for processing or they may be transferred to the main, long-term, memory (LTM). LTM has virtually unlimited capacity. Forging a path for retrieval of the symbols from LTM is done by repeatedly rehearsing them in STM, a process which takes a little time. Should STM capacity be exceeded because symbols are being read faster than they can be stored or because the subject tries to cram too much in, error is likely. The phenomenon of STM is linked with hearing. Possibly the evolutionary need for the memory resulted from the need for the temporary storage of spoken symbols. To interpret a spoken symbol in natural language requires accumulation of at least a few surrounding symbols to allow recognition of meaningful words or phrases in the stream.

Most people have STM capacity of between 5 and 9 random symbols. A random symbol here is equivalent to a symbol in a code (a letter, a digit) if all symbols are equally likely, i.e. the code is a random collection of symbols. However, a sequence of symbols in a code may form a meaningful collection. For example, they correspond to a previously-remembered string of symbols such as a word or year of birth or vehicle registration number. It would seem that such meaningful collections take up fewer STM places (perhaps because only a cross-reference to LTM is necessary). Assuming you are something of an English historian, you should find it a lot easier to remember the dates 1066 1492 1666 1812 than a random string of digits of the same length.

Although a large amount of literature on the capacity of STM exists, it may be a mistake to translate this too literally for application to clerical systems because the useful capacity of STM is probably affected by context-ual clues which abound outside laboratory experiments. For example, it is interesting that random codes of 12 symbols give rise to more transcription error than random codes of 13 or 14 symbols, and random codes of 9 symbols are worse than codes of 10, 11 or 13 symbols. However, it is difficult to be sure what is the number of random symbols that is equivalent to the non-random symbols that appear in a particular code. Error rates may also be affected by the learning, skill or experience of regular transcribers of codes. The following suggestions are consistent with experimental results and seem to be fair general rules for code design to minimise human error in recall and transcription.

1   Preserve meaning for the user. The transcriber should be aware of the semantic purpose of different parts of the code, if possible. Mnemonic symbols are preferable to cryptic symbols. Familiar codes may be preferred even if they belie an otherwise consistent rule.
2   Partition the code. Use hyphenation or spacing or meaningful facets to divide the code into at least two parts if it is 6 to 8 symbols, three parts if it is 9 to 11 symbols, four parts if it is 12 symbols or more.
3   Use alphabetic and numeric symbols. Do not intermix them randomly, i.e. keep all the alphabetic symbols consistently on the left side or the right side of the code, or, possibly, embed the numeric portion in an alphabetic prefix and suffix or vice versa. Thus AAA999, 999AAA, possibly AA999A or 99AAA9, but not AA9A99. (All-numeric codes may still be preferable for inexperienced keyers.)
4   Maintain consistent code length and format. If variation is essential, it should follow a simple rule. For example, home customers - customer code five digits, overseas customers - customer code six digits.

5   Aim to keep the code to 8 symbols or less unless the extra symbols very clearly add extra meaning to the code.
6   With codes that are used less frequently, abbreviation or encryption may be less tolerated and less necessary.
7   If the code is to be transcribed by the General Public, remember they may have no insights into facets of the code that are meaningful within the organisation.

When a code is transcribed by keying, immediate visual feedback aids error detection by unskilled operators. It has little effect on skilled operators. If unskilled operators are to key codes, consider restricting their search, e.g. by providing a numeric code which can be entered through a numeric keypad such as those found on calculators, or by using function buttons on a terminal.

People are likely to make reading mistakes with visually similar symbols, e.g. O,0,Q; I,1; C,G; Z,2. Also, presumably because of the aural origins of STM, similar **sounding** letters may cause confusion when being **read**, e.g. B,P; D,T; M,N; S,Z; F,V. Failure to detect correct spelling from spoken names is especially easy. The worst offenders here are the vowels and vowel groups, and silent letters like W and H.

Keying errors work out roughly as follows:

  50% substitution errors (wrong key depressed);
  20% errors of omission (the symbol is elided);
  10% adjacent symbol transposition (adjacent symbols reversed);
  20% other (non-adjacent transposition, format, content, insertion).

Questions

1   An application requires an unskilled operator to enter two 7-digit numbers via a numeric pad on a visual display unit. The two numbers are handwritten, separately, on a sheet of paper in front of the operator, who understands their meaning. A 'transmit' key is to be depressed when the operator intends the numbers to go to the computer. The computer program checks that it has received 14 numeric digits and rejects the entry if less than this are submitted or if any non-numeric symbols are present. If more than 14 symbols are sent, the first 14 are accepted and the rest ignored.
  Speculate on the proportion of errors undetected by the operator which will also be undetected by the program.   (5 min)

2   Speculate on the effect on errors (question 1 above) if the numbers are to be keyed by a skilled operator who is not aware of their meaning.
                                                                     (3 min)

3   Speculate on the improvement in error detection (question 1 above) that might be expected (a) if a symbol string in excess of 14 characters is rejected; (b) if the operator depresses the transmit key after each number, the program being amended to reject either number if it is not exactly 7 numeric digits.   (3 min)

4   On most computers, the digit zero has a different internal code and printing symbol from the letter O. Would it be feasible to give these two the same code and printing symbol?   (2 min)

10.2 CODES – ALTERNATIVES

A **sequence** code assigns successive numbers (or alphanumeric symbols) to successive items; for example, the cheque numbers in a cheque-book, numbered cloakroom tickets, the seat row letters in a theatre. Where the sequence code bears a temporal or spatial relationship to the things coded, as with the examples just given, it is not entirely cryptic. Arbitrarily allocated sequential codes, e.g.

    1 = Copper
    2 = Brass
    3 = Steel

are cryptic.

Deleted items will lead to gaps in a sequence code unless the deleted items are re-allocated. However, one must be careful with re-allocation, since the two different meanings of the code at different points of time may cause confusion. Also, if the new meaning actually obliterates the old meaning, as may be the case with computer-held records, there may be loss of information or even corruption of the meaning of old transactions.

A **block** code allocates arbitrary ranges of the available sequence numbers to different classes, e.g.

    001-200 Home supplier numbers
    201-999 Overseas supplier numbers

The blocks are cryptic. Within a block, the code is identical to a sequence number code. Unless the code values are an entirely stable set, the allocated ranges must include an allowance for future growth. This will lead to gaps in the sequence of numbers actually used (for possible consequence for file organisation decisions, see section 11.4). Changing a code structure used in a system can be a major operation, so all forseeable growth is usually generously allowed for. However, the possible short-term gains of having simpler, shorter codes should not be overlooked.

A **mnemonic** code uses some human association to relate the code to its meaning, e.g.

B = Brass	ENQ = Enquiry	HA	Home Accident
C = Copper	UPD = Update	HF =	Home Fire
S = Steel		FA =	Foreign Accident
	NTU = Not taken up	FF =	Foreign Fire
	OOS = Out of stock		
	ROP = Re-order placed		

Sometimes it can be desirable to change the uncoded word or phrase in order to find a consistent and easy set of mnemonics for a particular attribute.

A **faceted** code has different parts which represent different attributes ('facets') of the coded item. For example:

Theatre ticket: H7 = Row H, Seat 7.   Two sequence codes.

Glass: 6S = 6 mm sheet glass, 4F = 4 mm float glass.   A width attribute in millimetres, uncoded, with a mnemonic code.

Gas meter number: 1233241025011

digits 1-3: Locality (suburb/village). Sequence code.

digits 4-7: Walk number. Sequence code of streets, or part-streets, in meter-reading walking order.

digits 8-12: House number. Hierarchical code (see below).

digit 13: Tenancy sequence number. Incremented by 1 for each tenancy, dropping any carried digit.

Hierarchical codes are faceted codes where the meaning of a facet changes according to the value of another facet. Continuing from the example above:

House number: 02501

digits 1-3: House number (if any).

digits 4-5: Exception code.

              01 - No house number, so plot number given.
              02 - Four-digit house number. Add 1000.
              03 - First three are regular house number.
              X4 - The house number has a suffix letter, given by X.

The simplest and most maintainable system generally results from using only sequence and mnemonic codes. The other codes are harbouring a collection of attributes – it might be better to bring the attribute out into the open as an item in its own right. A register of issued codes should be available to users, either on hard copy or on a visual display.

Questions

1   Make lists of groups of alphabetic characters:
a)   whose **sounds**, in the context of spoken English, are alike;
b)   whose **sounds** in English are not regular;
c)   whose **names** in English (e.g. ESS is the name of 'S') **sound alike**;
d)   whose written upper case symbols are visually alike.   (20 min)

2   The income tax department has a file of all taxpayers on the computer. Taxpayers often call or telephone without a reference number. Sometimes a tax inspector hears a name and wants to check up on it. You are designing a system which will allow the tax officials to key in a name and initials in upper case and which will report back all the names (and other details) held by the computer which are equal to the input name. Also to be reported are all the names which **could** be equal allowing for the more likely mistakes in the name keyed. This could be done by having the computer form a coded version of the name according to the most reliable symbols keyed, and compare this with the stored names after they have been coded according to the same rules. State a possible set of rules.   (20 min)

3   The internal telephone system of a very large organisation is connected to a computer for abbreviated dialling of external calls. An internal call of format **5nn** connects to the computer, which uses the two digits **nn** to look up an external telephone number. For example, dialling 501 would tell the computer to get the second external number in the table (500 being the first). The computer sends the external number over the public telephone network, thereby connecting the caller. This allows callers to dial only three digits instead of the 8 to 10 digits that would otherwise be needed. Naturally, the numbers **nn** are going to be allocated to the most-frequently-used external numbers. A list of the allocations is going to be circulated to all internal telephone users. At present about 60 popular numbers have been identified. You have the task of making the initial allocation for the list. How would you do this?   (3 min)

10.3 CHECK DIGITS

A check digit is a suffix to a code which helps detect errors after the code has been transcribed. The most widely applicable system for an all-numeric code is the modulus-11 check digit, formed from the sum of the weighted digits in the code, taken to modulus 11. The check digit is calculated (usually on computer-produced lists) prior to the code being assigned, so that it becomes an integral part of the code in use. It is then checked every subsequent time that the check-digited code is input to the computer.
  The features of this system in outline are as follows.

1   Each digit in the code is originally counted in the sum after multiplying it by a weight. The weight varies according to the position of the digit in the code.
2   Considering a single substitution error in the code, where digit y_i is substituted for an unequal digit x_i (e.g. 256 instead of 246), and the weight is w_i,

$$(w_i y_i)_{\bmod 11} \neq (w_i x_i)_{\bmod 11}$$

where the integer weight, w_i, is in the range 1 to 10 inclusive. Thus all errors of substitution are detected.

3   Considering one transposition error in a code, digit x_i being transposed with digit x_j, there is no value of w (in the range 1 to 10) which will compensate for the changed sum if $w_i \neq w_j$ and $x_i \neq x_j$. That is, there is no value to satisfy the equation

$$(w_i x_j)_{\bmod 11} + (w_j x_i)_{\bmod 11} = (w_i x_i)_{\bmod 11} + (w_j x_j)_{\bmod 11}$$

Thus all errors of transposition are detected.

4   Considering all other cases, i.e. more than one error in a code, there are 11 possible values of the check digit, so if errors are going to combine by chance to create the correct digit, there are 10 ways they can fail and there is 1 way they can succeed. Thus 10/11ths of random multiple errors are detected.

For easy checking, the following procedure for check digit calculation is recommended.

1   Assign weights 2, 3, 4, ... to successive digits of the code in increasing significance (i.e. starting from the right).
2   Multiply each digit in the code by its weight — $w_1 x_1$, $w_2 x_2$, $w_3 x_3$, ...
3   Sum the weighted digits — $\Sigma w_i x_i$.
4   Divide by 11 and get the remainder — $\Sigma(w_i x_i)_{\bmod 11}$.
5   Subtract the remainder from 11, giving the check digit — $11 - \Sigma(w_i x_i)_{\bmod 11}$.

The check digit is then assigned to the least significant position in the code. To check a number formed in this way, it is only necessary to recalculate the weighted sum (including the check digit, which has a weight of 1) and divide by the modulus. If there is a remainder, the code is incorrect.

Worked example   Original code: 18472
                        Weights: 65432
                        Weighted digits: (6) (40) (16) (21) (4)
                        Summed weighted digits: 87
                        Summed weighted digits modulo 11: 10
                        Subtract from modulus: 11 – 10 = 1
                        Check-digited code: 184721

                        To check: 184721
                        Weights: 654321
                        Summed weighted digits: 88
                        Divide by 11: 8, remainder **zero**.

It will be seen from this procedure, at step 5, that if the remainder is 0 or 1, the check digit will be 10 or 11. If the check-digit would be 11, a check digit of 0 can be substituted instead. This still permits checking by the zero–remainder method as before. It is not usually convenient to have two–digit check digits, nor to substitute alphabetic codes. If the check digit would be 10, it is recommended that the code number giving rise to this check digit be discarded and another number used. This is easily arranged if the possible check–digited codes are produced by the computer on lists from which the next code number is drawn as codes are being allocated. For example, customer account numbers may be listed in the range of account numbers currently being allocated to new accounts. The codes giving rise to check digit 10 are simply omitted from the list. A similar method can be used if an interactive system is to supply the next account number, in response to a prompt. Some printing machines can pre–number printed stationery with check–digited numbers.
Should it be inconvenient to discard numbers, the best compromise is probably to use a modulus 10 system with prime–number weights (3, 5, 7, 11, ...). This will give a slightly lower detection rate on transposition and random errors.
Where a code contains an alphabetic portion, it is often decided to calculate the check digit on the numeric portion only. Alternatively, the

numeric equivalent of the letters (A = 1, B = 2, ...) can be taken and a check digit calculated using either a modulus of 11 (accepting a lower error detection rate) or a higher modulus, say 23 or 27, (leading to an alphabetic check digit to be added to the alphabetic portion of the code).

Questions

1   Account number 55436 has a modulus 11 check digit with weights as described in the text. Is it a valid number?   (5 min)

2   Assume the error rates mentioned in section 10.1 have been catalogued following examination of the 1% of codes which are erroneously entered (so 0.5% of codes contain a substitution error). Estimate the proportion of all codes that will have undetected errors, assuming it is a numeric code and a modulus 11 check digit is in operation.   (10 min)

10.4 USER CONTROLS OVER DATA INTEGRITY AND SECURITY

An important requirement of most systems is that the data stored by the computer should be of good integrity. By 'good integrity' is meant that all the desired data is accurately recorded, without omission, and stored on the computer safely, so that it is not accidentally or deliberately corrupted or lost. No spurious data should be accepted, and the recorded data should be internally consistent with one another. Analysts often loosely speak of this as 'security', but this term is also often more precisely used to mean prevention of unauthorised retrieval of data. Procedures which can be instituted in the computer department to maintain integrity and security will be discussed in chapter 14. This section is concerned with factors which should be the concern of 'users', treating the computer as a black box.
  The problem can be broken down into the following components:

1   messages which are to be input to the computer need to be recorded accurately in the first instance;
2   if a message is transcribed, it should be transcribed free from error;
3   all recorded messages should reach the computer or otherwise be accounted for, and no duplicate messages should reach the computer;
4   errors of original record or of transcription should be detected;
5   the entry of new data, or inspection of previously-stored data, should be done by authorised personnel only.

1   **Accurate original recording**   This is obviously crucial to everything else, but its importance is often overlooked. The main influences on accuracy of original record are probably as follows.

a)   The morale of the recorder (see section 8.2).
b)   Whether or not the recorder is involved with the accuracy of the data, by being involved with its purpose or by later having to rely on it. Broadly, whether he considers it 'his' data or 'their' data (see section 8.3).
c)   The training of the recorder in keying, if the data is to be keyed, or in writing if the data is to be written (see section 15.4).
d)   The design of the forms, displays and dialogues (see section 8.5 and chapter 9).
e)   The design of the codes (see sections 10.1 and 10.2).

In designing or reviewing procedures for recording data, a good question to explore is 'What are the **worst** mistakes that can be made?' Usually, mistakes in narratives such as names and descriptions are less important than mistakes in identifying codes, quantities, financial amounts. Particular emphasis should be placed on reducing the worst mistakes.
2   **Avoiding transcription errors**   Transcription errors are affected by the

following.

a)   Elimination of transcriptions by use of copies, machine-readable by-products, optical reading, turn-around documents, voice input, etc.
b)   The training of the transcriber.
c)   The clarity of the symbols used in the original message.
d)   The provision of immediate visual feedback, especially for an unskilled keyer, and facilities for the transcriber to correct his errors forthwith.
e)   The design of forms, displays, dialogues, codes.

3   No lost messages   Techniques include the following.

a)   Physical control. Spikes for documents, good filing and document handling equipment and routines, tidy work stations and work habits.
b)   Logs. Books or sheets to log transactions concluded, messages sent or received.
c)   Batches. Accumulation of documents into batches, which are then logged. The number of documents in the batch can be recorded separately, and the count checked on receipt. If important items in the batched documents are totalled, then checking this batch total on receipt can check for errors of transcription as well as omission (see 4 below).
d)   Sequence-numbering. If documents or other messages are sequence-numbered, as with cheques in a cheque-book, missing items in the sequence can be detected, as with uncleared cheques in a bank statement. Allowance must be made for spoilt documents if they are pre-numbered.
e)   Direct input. Keying data via a terminal, directly to the computer, removes the risk of loss except when there is a breakdown or some other interruption in mid-message. An acknowledgement of each complete entry submitted should be given, so that the terminal user is in no doubt whether or not his last message was accepted at the time of the breakdown. This still leaves the problem of the unreceived acknowledgement, i.e. the case where the computer gets the message and successfully uses it to update the database, but a breakdown occurs before the acknowledgement is sent to the terminal. The operating system may permit undoing of the transaction or queueing up of the acknowledgement for re-transmission when service is reconnected. In any case, provision should at least be made to allow a user to discover the status of his accepted and non-accepted entries at his last session.

4   Detection of origination or transcription errors   Possibilities include the following.

a)   Sight-checking by an independent person.
b)   Asking (e.g. in a dialogue) for repetition or confirmation of 'worst mistake' items. Repetition may be indirect, e.g. asking for age as well as date-of-birth.
c)   Verification (independently re-keying the whole message). This may be called for when uninvolved transcription, with no visual feedback, is practised.
d)   Validation by the computer, providing rejections of impossible messages, warnings about unlikely messages, and proof lists of accepted messages.

Validation techniques leading to rejection include:

a)   Range checking, i.e. checking that the value of an item is in a permissable range. If customer numbers are presently in the range 1000 to 2982 and all forseeable growth is catered for by the range 1000 to 3999, reject any customer numbers below 1000 or over 3999.
b)   Class checks, i.e. checking the value of an item is correctly alphabetic or numeric, or that the symbols in a code have the correct class given their position. For example, the customer number in a) above must be numeric. (It is not enough just to have a range check here, since a middle symbol

may be alphabetic but programs may still consider the customer number in range, e.g. customer number 29A1.)

c) Format checks, i.e. checking that the predefined format rules are followed if the message is to contain identifying symbols or special characters such as leading or trailing arithmetic sign, currency sign, punctuation. (Also consider allowing choice over format; see sections 8.4 and 8.5.)

d) Internal consistency checks, i.e. checking that related values in the message are consistent with one another. In this category falls check digit validation (is the check digit consistent with the rest of the code?) and repetition validation (is the second entry the same as the first?). Other consistency checks should be considered for the particular case. An example could be: if 'age' attribute has value less than 16, 'driving licence number' attribute must be absent or blank.

A **hash total** is sometimes added to a message to permit a special internal consistency check of the 'worst mistake' items. At source, designated numeric items are added together (even though it is meaningless to do so, e.g. dates added to quantities) and this 'hash' total is recorded at the end of the message. The computer re-calculates the hash total and compares its results with the original. A transcription error has occurred if they disagree.

e) Preceding message consistency check, i.e. checking that the values in the present message are consistent with the values in a preceding message or messages. Perhaps messages must appear in a certain sequence (deletion of a customer record may follow an amendment of that customer, but not vice versa) or duplicate identifiers are not allowed. Batch total validation ensures that the batch count or amount total, or possibly a batch 'hash' total, is correct.

f) Database consistency check, i.e. the values in the message are checked for consistency with data held on the database. For example, a customer exists for the change-of-address submitted, a Caesarian section is recorded for female patients only.

Instead of rejecting the message, a warning may be given that a particular value is unlikely; for example, very large, but not impossible, amounts. Usually such warnings are passed and then forgotten by the computer system. The onus is then on the user to check out the case and take corrective action if necessary. A more positive possibility is to have the system 'remember' all warnings until either an amending entry or a confirmation of the original entry is received.

In some circumstances, it may be desirable to treat even clear errors as having only warning status, perhaps because the seriousness of the error does not outweigh the urgency of processing the rest of the message. Such a case may occur with errors in classifying codes which are used for management statistics at month end, but which do not prevent processing of important orders in a batch system. These cases should be remembered by the system until a correcting entry is received. Any outstanding items should be listed out periodically for chasing, and the list should be clear before using data for the affected purpose. Alternatively, the erroneous data can be used and the management reports interpreted in the light of the outstanding errors. The erroneous items can still be corrected later, but this may not be satisfactory if the statistics are important for the analysis of a time series. (The correction should be made by contra-entry and resubmission – see section 14.3.)

Proof lists are printouts of all the accepted messages for immediate visual checking, or for filing for subsequent detailed checking should it be suspected that an error exists. In a batch system, it can be checked that the difference between the accepted message totals and the original batch totals is accounted for by the rejected messages. Of course, the list of rejected items can be used as a basis for checking that all corrections have been submitted. A system which exercised more positive control over the corrections would have the computer remember all rejected messages and tick them off as corrections are submitted, printing out outstanding rejections periodically, for chasing. Lists of present master file contents are also useful for visual checking and error-tracing.

5   **Security**   Attention should be given to possible needs for the following.

a)   Physical security. Means of access to department or terminal room. Locks on doors, files, terminals. Guards.

b)   Supervision. Responsible scrutiny of activities. Requirements to obtain authorisations. Procedures for checking authorisations. Provision of monitoring information on a supervisor's terminal – logons, logoffs, logon attempts that fail, transactions being made.

c)   Passwords and usercodes. Access to computer restricted to those persons who identify themselves with a valid usercode and give the right password or supply a valid badge. Limited facilities available to a usercode, e.g. enter data only, interrogate only, compile programs. Limited portions of the database or data elements which may be accessed, e.g. read personnel data but not salaries. Passwords should be non-printing or erased straight away after entry, subjected to occasional change, and withdrawn from the computer when authorisation is withdrawn from the user.

d)   Automatic logoff. If a terminal sends no message within a prescribed time, require re-logging of the user.

e)   Suppression, obliteration, encryption or non-reproduction of sensitive information on documents or copies sent out.

On the whole, computerised systems may, even without any special steps being taken, give greater security than exists in the equivalent manual systems. Even with very-high-security systems, there may be much more risk of unauthorised use of data by persons who are authorised to access it, than unauthorised access.

Fraud or embezzlement with computerised systems is also perhaps not quite such a likely possibility as would appear from the amount of attention given to it (Cromer, 1). The most common types of fraud entail the input of false transactions, e.g. a purchase order placed on a non-existent supplier. The system of authorities should be arranged so that collusion is necessary to enter a fraudulent transaction, e.g. a new supplier record can be created only with the manager's authorisation, and purchase orders can be raised only by the purchasing clerk.

Theft by illicit code in programs, etc., while not entirely fanciful, is also rather unlikely. With most white-collar crime, the most powerful deterrent is the fear of detection and exposure; it should be widely known that spot-checks of data and scrutiny of programs, etc. is undertaken.

Questions

1   An ideal error-detection system will also (a) pin-point the place of the error and (b) permit easy correction. Use these criteria to compare the following systems.

A   Orders are taken down on written order forms by clerks. The clerks batch the forms in tens and add up the batch total of 'quantity ordered', which is entered on a batch header document. The batch number is logged and the batch sent for keying and verification, and validation by the computer. The computer sends back a list of rejected batches and a list of rejected orders within the accepted batches.

B   Orders are entered by clerks onto a 'form-filling' display on a VDU terminal. The computer, which has master files online, validates the display when the transmit button is pressed.

C   Orders are entered by clerks onto an intelligent key-to-disk terminal with a one-line visual display. The terminal validates the data as each field is entered, and sends valid orders to a disk. The disk is later batch-processed by the computer.   (10 min)

2   The claim in the text that a transcription error has occurred if a hash total fails is not quite true. Why?   (2 min)

ASSIGNMENT

Continuing the assignment at the end of chapter 9, design the codes for the mail-order wine club. Suppose the founders have asked that the wine numbers be kept to two digits, to keep things simple; should you go along with this? Complete the definition of the business procedures for their proposed system. Conclude the documentation of these procedures (not computer files or programs) and prepare:

a)   System Outline and System Flowchart;
b)   Clerical Procedure Flowcharts as necessary;
c)   Document and Display Layouts, togther with Clerical and Computer Document Specifications, Record Specifications for display/print lines if needed, and a data dictionary.

 Should you be working in a pair, a possible allocation of work would be as follows.
Person A: Order form design, order entry displays/dialogues, order entry procedures, system outline, codes design.
Person B: Order confirmation design, picking list design, picking and despatch procedures, system flowchart, data dictionary.
Together: Review documentation.

ANSWER POINTERS

Section 10.1

1   A substitution of non-numeric symbols may be considered unlikely when a numeric pad is used, so substitution errors will be numeric substitutes, undetected. These account for about half the errors.
 Omissions will all be detected. These account for about 20% of errors.
 Transpositions and insertions will not be detected, say 15% of errors.
 Format errors will be detected, but not content error if the format is correct.
 Speculatively, about two-thirds of the errors undetected by the operator will remain undetected by the program.

2   Loss of meaning may be a red herring for the stated task, since two 7-digit numbers are probably rather meaningless (unless they happen to be the user's home and work telephone numbers). Unskilled operators using a full keyboard can have error rates easily ten times that of skilled operators. Presumably, differences are less when a numeric pad is involved. Speculatively, for this task, error rate reduced a little.

3   (a) This will lead to errors of insertion being detected. Marginal improvement, maybe, at most a few percent.
(b) Assuming that the operator mentally picked up each of the two numbers separately in the first method, improvement just from partitioning is not likely. However, if this means that the display is now partitioned, when it was not partitioned previously, this may help the operator to recognise errors which have been perpetrated (particularly in the case of an unskilled operator). Also the increased uniformity of action and fulfilment of a natural expectation that some action should distinguish the first number from the second could improve accuracy.

4   Yes, and this would eliminate zero/oh confusion. A problem might arise if it was desired to have the computer translate upper case characters into lower case. The code would not conform to international standards.

Section 10.2

1   There is no definite answer (how alike is 'alike'?). A possibility:
a) B,P; D,T; M,N; S,Z,X, soft C; K,Q, hard C; F,V; J, soft G. Also vowels.
b) Vowels, silent or aspirate H, hard/soft C, G, S, and many other irregularities, silent letters.
c) B,P; C, Zee (US); D,T; M,N; I,Y.
d) C,G; M,N; O,Q; U,V. A lot depends on the characteristics of the typeface or handwriting.

2   Possible scheme: eliminate the vowels, the near vowels Y and W and the unreliable H. Then code source and target names according to similarity groups by chaining all confusions, e.g.

    B,P = 1
    C,G,J,S,X,Z,K,Q = 2
    D,T = 3
    F,V = 4
    M,N = 5
    L = 6
    R = 7

H, W and Y, etc., are more reliable as beginning letters in English, so a possible code would prefix the beginning letter, then translate. So:

    PARKIN P725
    PARKINSON P72525
    HOWE H
    HOUGH H2
    SMITH S53

The examples illustrate that it could be desirable to take into account some of the more common dipthongs, such as GH. Regional accents and immigrant populations make generalisation difficult. It might be sensible to consider a pilot scheme as above, with the first one or two initials also forming part of the search key, but it would be best to parameterise the groups to allow experimentation and refinement locally in the light of results.

3   If frequency-of-use of the numbers is skewly distributed, the easily-remembered numbers 500, 555, 599, then 511, 522, etc., could be allocated to the most popular cases. The list could be sequenced alphabetically. There may be a small gain, if use frequency is very skewly distributed, in repeating the most-frequently-used items at the head of the list.

Section 10.3

1   Weighted sum = 68. 68 modulo 11 = 2, so not valid.

2   Assuming that all errors of format, insertion, and omission are detected by other checks, this leaves errors of substitution, transposition and 'content'. 'Wrong content' is presumably equivalent to 'multiple errors'. Altogether about 0.7% of the codes.
 All errors of substitution (0.5%), transposition (adjacent and non-adjacent - say 0.15%) will be found, leaving 0.05%. Ten elevenths of these errors will be detected, leaving about 0.005% undetected; 5 errors per 10 000. If wrong content could be picked up from a legitimate source, though, this would tend to defeat the checking.

Section 10.4

1   If a batch control total fails, the error may be in any one 'of the documents in the batch, or an omitted document, or the batch total. In this respect, B and C pin-point errors better. With suitable error message design, all three can pin-point other errors.

 System A is least easy for corrections because of the time lag, retrieval of source material and the need to resubmit fields which were not in error. System C also has a time lag for database inconsistency errors. Other differences between B and C depend on where the cursor is left and details of the error correction and re-transmission procedures.

3   The hash total may have been incorrectly calculated.

REFERENCES

(1) Cromer, M., **Computer Fraud**, The Law Society's Gazette, 10th January, 1979, pp. 14-15.
(2) Eason, et al., op. cit. end of chapter 8.

11 Database definition

11.1 HIGH–LEVEL DECISIONS

Defining the database entails defining a database 'schema', or the files and records which are to form the basis of the computer part of the system. Database design is an art, not a science, and even if a complete set of rules can be given, it is hardly practical to expect an analyst to follow them. The objective of this chapter is to give a practical method which will result in an initial simple design. The analyst can subsequently tune this if he chooses, although tuning quite often brings increased complexity. This book is concerned only with high–level decisions about database content, file organisation and access. More detailed decisions are considered to be 'programming' decisions. This is not to suggest that such decisions should not be made by analysts (nor, for that matter, by users); only that such decisions are considered a separate topic, outside the present scope.

Even when a data model has been well drawn up in the first instance, it is likely that additional attributes will be identified when attention is turned to the computer procedures. (It is to be hoped that this will not involve any additional entities or relationships.) These attributes are likely to be status flags and control fields which arise out of the desired computer procedures. The procedures also dictate the desired access paths to the data, which in turn affect file organisation. The distinction between designing data structures and designing procedures, preserved here for the sake of this explanation, is still artificial; the analyst will need to alternate his attention between the two. Systems analysts find, with experience, that they can make quite good decisions in advance about the database, even at their first attempt at definition. However, rigorous subsequent scrutiny of the details is needed, since loose ends in the analysis may be difficult to tie up later.

If the data model has not been well drawn up, database design may not be so straightforward. Experienced analysts can often create ad hoc order for the database out of quite chaotic information about the attributes to be recorded and the procedures desired. It is preferable, for a system which is to be used often, to go back and revise the data model. This may have repercussions on the business procedures which have been specified, but this risk does not outweigh the need to define the business procedures before the computer procedures.

Broadly, the different importance attached to data analysis in the first instance can lead to different styles of system development. If data analysis is emphasised, analysis and design stages tend to be prolonged but construction and trial phases tend to be short. If data analysis is played down or omitted, analysis and design tend to be quickly completed, but construction at least runs the risk of being more lengthy and trials are more likely to disclose needs for corrective action.

The explanation that follows is in the context of a design that does not use a Data Base Management System (DBMS). However, the philosophy of the approach is the same, whether or not a DBMS is used. With a DBMS, much discipline is imposed by the system and associated conventions. Without a DBMS, the systems analysts need to impose their own discipline.

Question

1   To what extent is it true that attention to data analysis will bring worthwhile returns from the construction, trial and operation phases?
(10 min)

11.2 TYPES OF FILE AND OUTLINE METHOD

The files, and their records, which are to be designed will be of the following types (these definitions may differ slightly from popular usage).

1   **Master files and transaction files**   These collectively contain all the data about the entities and relationships in the data model. Other files that may exist contain only copies or derivations of master and transaction file data.

2   **Master files**   A master file is one which is not normally emptied of data. In the normal course, a master file is the preserved file that is the product of successive updatings of an originally-created master file. It may have its entire content changed as a result of being updated, or it may remain unchanged throughout its life. The **current** master is the one which has most recently been updated. Examples: customers' names and addresses and account balances; suppliers' names, addresses and account numbers; table of department codes and department names.

3   **Transaction files**   A transaction file is one which accumulates records and which, in the normal course, is cleared of the records at arbitrary points of time. Usually, the 'arbitrary points' are when the transaction file has been used to update a master file. After such an updating, the transaction file records are removed and the file is re-accumulated. Often, 'removal' of the transaction file records is achieved by removing the physical medium which contains the file. In this case, a blank medium, or medium containing an empty edition of the file, is substituted for the superseded version, to begin accumulating further transactions. The **current** transaction file is the one presently being used to accumulate records. Examples: orders; stock issues; deliveries; name and address changes; payroll changes.

4   **Workfiles (intermediate or programmer's files)**   Workfiles are the other files needed to support the computer procedures. They contain only data values which are copies of those already stored in (or perhaps yet to be stored in) the master and transaction files. Sometimes workfiles are just exact copies of master or transaction files in a different sequence (in which case one can debate which copy is the workfile and which the original, but it doesn't matter). Workfiles may be designed by the systems analyst to support the computer procedures. Some workfiles may be created auto- matically by software, e.g. the temporary files created by a sort utility. Others may be created ad hoc by a programmer to solve a programming problem. In a sense **all** workfiles are programmer's files, because they exist to serve the 'programming' need: the need temporarily to reorganise the data (e.g. by sorting, merging, reformatting, extracting, summarising) in order to give a new data structure which will permit simpler algorithms in the computer programs, which will overcome limitations in the computer hardware or software, or which will permit more efficient processing by the computer.
 If all the data in the master and transaction files were accessible by all required keys, if all the new transactions were subject to online validation, if there were no need to husband the computer resources or consider the simplicity of algorithms, then there would be little need for workfiles.

5   **Security files**   Security files are copies of or derivations of master or transaction files, not used in the ordinary course of processing, which allow reconstruction in the event of loss or corruption of data, or which permit

the accuracy of the data, and the correctness of the operation of the computer procedures, to be established. Workfiles or superseded versions of master and transaction files are often used to fulfil this role. Discussion of security files is postponed to chapter 14.

The following is a practical method of designing computer files and procedures.

1   Take stock of constraints   Constraints may be imposed by hardware or software or by data processing policies. The constraints may limit device availability, the type of file organisation or access method which can be used, the software utilities or packages which can be used. Other things being equal, one does not want to design a system around a device that is not available or not supported by the mainframe, nor around a file organisation that is not supported by the software. Nor does one want a design that contradicts the policies of the department. In practice, the constraints make the high-level design decisions easier by removing some of the possibilities. Of course, constraints are often relative and one must avoid the trap of accepting a constraint when in fact it would be a net advantage to buy new hardware, develop in-house support for a file organisation or revise policy. I shall limit the discussion to three common file organisations later (sequential, relative and indexed – the only ones supported by standard COBOL), since to consider all possibilities is hardly realistic.

2   Take stock of priorities   Particularly from the point of view of the trade-off between firstly, the investment of analysts' and programmers' time and secondly, the efficient use of computer resources. The relative decline of computer memory and backing-storage costs, together with the increase in analysis and programming costs, continues to tip the scales towards ease of design, programming and testing at the expense of hardware time and capacity used. It may be cheaper to buy extra hardware or software if this simplifies development. The possible purchase of a package should not be overlooked. Sometimes, short-term considerations act to distort the priorities (e.g. systems analysts' and programmers' salaries have been budgetted for, but hardware enhancements have not). Only forward planning or an alert and flexible management can combat this.

3   Define the content and layout of records   The records of master files and transaction files should be defined; this means taking the attributes of the data model and organising them into record types and files – see section 11.3.

4   Outline-plan computer procedures   Set down the procedures that are going to use the master and transaction files (ignoring at this stage the possibility of introducing workfiles). This is a back-of-an-envelope design to assess what types of access will be made on the master and transaction files and how frequently they are processed – see section 12.1.

5   Choose the file organisation   For master and transaction files – see section 11.4.

6   Design computer runs   This also means specifying the alternate indexes of master (and possibly transaction) files, the access method to be used by programs (search, skip-sequential, direct), design of workfiles and security files (their content, layout, organisation etc. – see chapter 14). This may lead to the addition of security control records or attributes in the master and transaction files, if these have not previously been attended to. Computer run design is outlined in section 12.1.

7   Specify computer programs   See section 12.2. This step may in practice entail making many detailed decisions such as device allocation, blocksize, overflow areas, number of keys per index entry in coarse and fine indexes, buffering, file geographies, etc. Most of these are considered in this book to be programming decisions, but some are covered in chapter 13.

8   Review earlier steps and finalise documentation   See section 12.3.

Question

1   Standard (full ANS) COBOL allows indexed files to be defined with an alternative index or indexes, in addition to the primary key. The alternate index may have duplicates if desired. How does this help eliminate workfiles as defined above?   (2 min)

11.3 FILE AND RECORD CONTENT, RECORD LAYOUT

The attributes of the entities and relationships can be translated into record layouts by a process of **recombination** and **splitting**. The layouts are refined to cater for **optionality** and then grouped into **files.** What we are doing is 'flexing' the data model (chapter 7) to explore the effect of alternative representations on the computer.

1   Recombination
Start out with the assumption that each entity-type and each relationship-type will give rise to a type of record. Considering a pair of entities joined by a relationship giving rise, at face value, to three record types.

a)   If occurrences in the two entity-types have a relationship of degree one-to-one, merge the attributes of the two entity-types and their relationship into one record. (Probably this case will not arise, because it is likely it will have been conceived as one entity-type in the first instance.)

b)   If the degree of the relationship is one-to-many, merge the attributes of the relationship into the entity on the 'many' side, making two record-types, one for the 'one' entity and one for the 'many' entity. So if a relationship has no attributes other than the primary keys of the entities concerned, this operation will simply post the primary key of the 'one' side into the attributes of the 'many' side. Similarly, if the relationship has other attributes, the 'many' entity will pick up the primary key of its related 'one' entity and the attributes of the relationship it enjoys with the 'one' entity. (Again, there might be no relationship like this with attributes, since such attributes may have been seen as attributes of the 'many' entity anyway.)

c)   If the degree of the relationship is many-to-many, consider the possibility that a one-to-many relationship is adequate for the anticipated procedures. For example, a many-to-many relationship may arise as a succession of one-to-many relationships over time. If only the **current** attachment of the employee is to be the concern of a formal message from the computer, then the many-to-many employee-department relationship may be treated as many-to-one. The 'current department' key would be posted in the employee record. With this structure, we can expect to answer easily the questions 'Which employees are in department X?' and 'What is the current department of employee Y?', but it will be difficult or impossible to answer a question 'Which departments has employee Y ever been in?' or 'Which employees have ever been attached to department X?'
  Even when the latter type of question is to be asked, it may still be possible to treat the relationship as one-to-many on the **computer** system. If details of what happened in the past are simply a copy of states which at the time were reflected in a 'current' state, if a hard-copy record of that current state was created and preserved, if all the queries can be satisfied by the inspection of the preserved hard copies, then only the 'current' part of the data model need be reflected in the computer system (ignoring, for the moment, recovery requirements). The 'past' part of the data model is realised by the hard copies, and the questions are answered by the **manual** system.
  Failing simplification into a one-to-many relationship, it may suffice for

a stable, fixed number of entities on one side to be permitted to participate with a single entity on the other side. For example, if only the current and immediately previous departments of an employee are to be the concern of formal messages from the computer, the employee–department relationship can be treated as many-to-two. In this case, two lots of department attributes would be provided in the employee record.

If it is still considered that a many-to-many relationship must be recorded (the safest conclusion), then the simplest and most flexible choice is to preserve three record types. For example, there would be one record for each employee, one record for each department and one record for each employee–department relationship that has occurred.

If more than two entity-types are joined by a relationship, then the record design can proceed in similar fashion. For example, if the relationship is one:one:many, the keys of the two 'one' entities can be posted in the 'many'. A clear head must be kept when dealing with many:many:one relationships or those of even higher degree, because such relationships are ambiguous (it will probably help to reconceive the relationship as an entity and consider the relationships that pair this new entity to the previous entities). If in doubt, the solution that will always **work** (and will always be simple and flexible) is the one that keeps separate records for each entity and relationship.

2   Splitting for efficiency

Splitting is redesigning a record into two records by separating out some of the attributes if these attributes are not required every time the original record is accessed. The purpose is to reduce the transfer of unwanted data to and from backing storage. Splitting is not essential to a working solution (although failure to split may strike up against the constraint of available processing time), but it is often desirable for more efficient processing of master files. A record is divided into two or more records, one of which is used for one purpose or set of purposes, the other being for other purposes. There is a penalty to pay in the total amount of backing storage used, since the primary key of the original record must be repeated in the split versions. An example is as follows.

Original record: COMMODITY MASTER (COMM–NO, QTY–PER–PACK, PACK–PRICE, PACK–POINTS, DESCRIPTION, SALES (24 CUSTOMER GROUPS (LAST 8 WEEKS, LAST 24 MONTHS)))
Possible split record: COMMODITY (COMM–NO, QTY–PER–PACK, PACK–PRICE, PACK–POINTS, DESCRIPTION): SALES ANALYSIS (COMM–NO, SALES (24 CUSTOMER GROUPS (LAST 8 WEEKS, LAST 24 MONTHS)))

In this example, it is assumed that a gain is won by the ability to process daily delivery notes using COMMODITY only. The complete data about commodities, required say during weekly invoice production, can be obtained from the combined records. Of course, additions to or deletions from the files split in this manner must be carried out in harmony on both sides.

Sometimes this division may permit a further efficiency gain by allowing the split records to be normally physically stored in differing sequences or with different organisations. Duplication of non–primary–key attributes in the split records should be avoided, since this leads to two versions of the same attribute, with potential difficulty if the attribute values stored are ever at variance with one another as a result of one record of the pair being updated when the other is not. Such duplication can be the source of efficiency gain by removing a need to consider the combined records, but the gain may be hard won. If attributes are duplicated, the procedures should make certain that all versions are updated whenever a change occurs. Human–originated updates for all the alternative versions are not reliable, so there should be computer procedures to do the updating. These procedures need to be controlled for correct operation, just as with other procedures.

The increased complexity may outweigh the original gain.

There is no point in splitting files if there is little clear gain, e.g. the file is of small size and is infrequently used.

3 Optionality

Where participation in a relationship is optional, this is automatically reflected in the state of the files if separate records exist for each entity and relationship. Where recombination has been done, the posted relationship attributes must be either omitted or null when the entity concerned does not participate in the relationship. 'Null' in this context usually means filled with zeros or blanks, according to the class of data. Similarly, other attributes of the entity, or groups of attributes, may be omitted or null if they are optional.

Omitting optional attributes will mean the record type has variable length. Variable length records may give rise to a saving in file size (especially if there are many optional attributes which have long values, and the options are infrequently used). The extent of any increased complexity caused by variable-length records, and the amount of the saving in file size – there may be none – is dependent on the supplied software. A big saving should be produced before increased complexity is justified.

Analysts sometimes design records which contain an attribute or attribute group which optionally can be repeated an indefinite number of times. There may be complications in dealing with continuation records if there is a limit to the maximum physical record size. Such cases will not arise if the data model was fully normalised, since the repeating attribute(s) will be separate entities or relationships deserving a separate, fixed-length record.

Supplying null values preserves fixed length when a record contains optional attributes. A reduction of storage requirements might be achieved by splitting the records into two or more record types, each of fixed length. One record type would contain the mandatory attributes and the frequently-used, short options or optional groups. The other(s) would contain the long options which are infrequently taken up. Of course, there is a penalty in adding the primary key to the split-off record. For example,

> Original record: DRIVER (LICENCENO, PERMITTED-CLASSES (4), LICENCE-FEE, DATE-OF-BIRTH, DATE-OF-RENEWAL, DISABILITY (DISABILITY-TYPE, DISABILITY-CERTIFICATE-NO, DATE-OF-EXAMINATION, NAME-OF-EXAMINER) ...)

Suppose that permitted-classes 2 to 4 are optional, but the values are short codes and one or two options are often taken up. The disability attributes are an optional group, rarely used, with long attribute values. The record is thus divided as follows.

> Divided record: DRIVER (LICENCENO, PERMITTED-CLASSES (4), LICENCE-FEE, DATE-OF-BIRTH, DATE-OF-RENEWAL, ...); DISABILITY (LICENCENO, DISABILITY-TYPE, DISABILITY-CERTIFICATE-NO, DATE-OF-EXAMINATION, NAME-OF-EXAMINER)

Note that there is an additional complication resulting from division-for-optionality that does not occur with splitting-for-efficiency as previously described. In the previous case, there was a one-to-one correspondence between all split records (every commodity had both a description and a sales record). In the optionality case, there is no such correspondence (not every driver has a disability record). This may make the rules for processing, updating, adding or deleting records more complicated. Some simplification may be possible by inserting in the mandatory record a flag which indicates the presence of the optional record.

Consideration should be given to adding unused fields to records in order to cater for possible future expansion or unforeseen requirements. Analysts often allow 10–30% of record length, depending on how uncertain they feel about the design. (But the operating system may permit change of record length in a file without recompilation of programs using the file; in this case, there is less need to allow for expansion at the end of records.)

4 Files

Starting off with the assumption that a file will exist for each record type, the only remaining issues are whether or not the records that result from splitting should remain in separate files or be placed in one file, or whether or not records for different entities or relationships should be mixed in the same file. Sometimes, splitting-for-optionality results in an optional record very similar to the mandatory record. Such records could be placed in the same file without significant increase in complexity, especially if they correspond to the idea of 'continuation'. In other cases, the separate-file solution is generally the easiest and the most flexible. Unfortunately, it may not always be implementable because of constraints of the computer system, e.g. limited number of files online at one time.

If split record types are assigned to the same file because of constraints of the computer system, there may still be efficiency gains of divided records if the two record types can be selectively accessed within the file. Failing this, the only efficiency gain likely to be of substance is in diminished record area size in main memory; however this is often only a small gain and not very valuable on large modern computers. On a very small computer, or with very large records, record area size might be a constraint. If this gain is not thought worthwhile though, the records might as well be recombined. (There may be a further slight gain if maintaining fixed-length split records avoids the need to have length counts for variable length records. This gain will be negligible if the records are at all long, and the absolute gain is likely to be small anyway.)

Mixed record types, not resulting from division, often result in a loss of efficiency. They also increase complexity. Nevertheless, a lot of constrained systems are designed in this fashion, e.g. employee records embedded within department header records, invoice records embedded in the customer accounts file.

As a result of these four steps, the file and record layouts will be drawn from those illustrated in Figure 11.1. If the simplest, separate-file choices are made throughout, all records will be of type (a), (b) or (c), but there may be many files.

Questions

1   From the recombination discussion on page 138, if an employee-department relationship is many-to-two, what questions can one expect to answer easily in addition to 'Which employees are in department X?' and 'What is the department of employee Y?'.   (4 min)

2   Suppose EMPNO has length x characters and the employee record, a candidate for splitting for efficiency, has other payroll data with length 150 characters and other personnel data with length 100 characters. The split-off payroll file PAYDATA would be passed once each month for the payroll. The two files, PAYDATA and PERSDATA, would be passed together 8 times per month for updates and enquiries. Is there a length of EMPNO which would make it worthwhile (from the point of view of the total monthly characters transferred) to recombine the records?   (15 min)

3   (a) Continuing from question 2 above, suppose EMPNO is 7 characters but it is not known how many times update and enquiry passes will be required. Is there a frequency of use of the combined files which would make it worthwhile not to split the records?
(b) Can you propose a general formula for deciding whether or not to split records for efficiency?   (15 min)

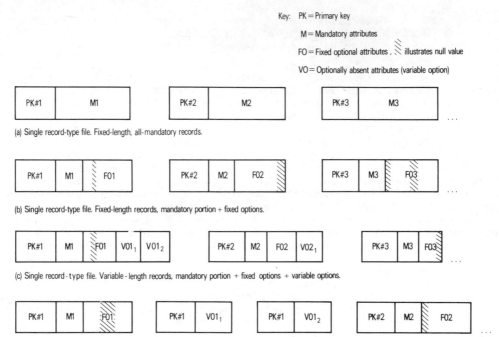

Key: PK = Primary key

M = Mandatory attributes

FO = Fixed optional attributes , ⧄ illustrates null value

VO = Optionally absent attributes (variable option)

(a) Single record-type file. Fixed-length, all-mandatory records.

(b) Single record-type file. Fixed-length records, mandatory portion + fixed options.

(c) Single record-type file. Variable-length records, mandatory portion + fixed options + variable options.

(d) Multiple record-type file, illustrating a mandatory portion + fixed options record type and two optional record types. Records of a given type are fixed length in this example, but the lengths of different types of record are unequal. Padding-out can make all records of equal length should this be desired.

If physical contiguity on the medium is reliable, the repeated Primary Keys in the optional records may be dropped; blocking will then transform this example to case (c) above (ignoring housekeeping fields). Separately filing the records of this example will lead to one file of case (b) -with every Primary Key present - and two files of case (a).

Fig. 11.1: Types of file and record

4   A file is to contain the following data items:

```
CUSTOMERNO   7 characters
ORDERNO      5
ORDERDATE    6
LINENUMBER   0 or 1 (see text)    ) repeated
PRODUCTNO    3                    ) for
QUANTITY     3                    ) each
VALUE        6                    ) line
```

The following estimates have been made:

Number of order lines per order	% of orders	Weighted average
1	5	.05
2	50	1.00
3	30	.90
4	10	.40
5	4	.20
over 5	1	.05 est
		2.60

A maximum of 10 lines is allowed per order. There will be an average of 1500 customer orders on the file.

Estimate the file size for the following cases.

(a) The records are fixed length, permitting 10 order lines (the line number is zero length in this case since it can be determined from physical position in the record). (Like Figure 11.1(b).)
(b) The records are of variable length and an additional three characters are needed per record for housekeeping fields (line number zero length). (Like Figure 11.1(c) but with no fixed options.)
(c) The records are of fixed length, one order line per record, the line number being added to the key of the record. (Like Figure 11.1(a).)
(d) The records are of fixed length, but contain 2 order lines per record. Orders with more that the set number of lines will carry over onto one or more continuation records with the same layout. A one-digit continuation counter is added to the key – this removes the need for line numbers. (Like Figure 11.1(b) again, but resulting from division of case (a) above for optionality.)
(e) As (d), with 3 order lines per record.
(f) As (d), with 4 order lines per record.   (30 min)

11.4 FILE ORGANISATION AND ACCESS

This section assumes that sequential, relative and indexed organisations (as defined in standard COBOL) are available. Initiative must be applied to adapt the advice to more or less constrained cases.

  Choose relative organisation for any files which have no gaps in the primary key sequence and where a record exists for every (or nearly every) possible primary key. This means that the file will be self-indexed by the primary key, or by the primary key after it has been transformed by some simple rule such as the addition or subtraction of a constant. In practice, usually only a few master files or tables satisfy this requirement. If no direct access to the files is required, choose sequential organisation for the remaining files.

  This advice leaves us to consider the question 'Is direct access required?' If the entire file is to be processed by each program referencing it, with each record being accessed in turn but not selectively, and without records being added other than at the end of the file, direct access is not required. This is often the case with transaction files, or can be arranged to be the case with such files by choosing a suitable file sequence. In other cases, direct access to retrieve, insert or delete a record of nominated key is always desired at first sight, since it leads to simpler procedures and removes the need for workfiles, sorting, searching, and written-forward copies. The problem is that on backing-storage devices where mechanical movement is needed (e.g. seeking and rotation on disks), there may be an intolerable penalty if direct access procedures call for much more of this movement than would be the case with search access procedures. This penalty may outweigh the penalty of transferring unwanted records in the search. There may also be a penalty of increased space required for storing a direct access file, but this is often more tolerable.

  In the case where a large number of programs are referencing files on the same backing-storage device, and the operating system queues up many input-ouput requests to satisfy them all in one complete traverse of the storage on the device, there is virtually no time penalty for direct access from the viewpoint of an individual program, so direct access is 'required'. (An exception may occur if index and overflow retrievals lead to a much greater number of accesses than would occur with search access – such cases are not so clear.) Quite a reliable rule with such an operating system is 'sequential organisation for transactions, indexed organisation for masters', but this is not infallible. (It may be worthwhile, of course, to access an indexed master skip-sequentially if the hit rate is high. This can be decided by quantifying record and index accesses for the specific program and discovering which access method results in the least input-output processing time. If skip-sequential access means that transactions must be sorted, the sort accesses must be added to the skip-sequential accesses in making this comparison, a consequence which tends to tip the balance in favour of unsequenced direct access again.)

In the remaining cases (in which a penalty is incurred with direct access), direct access may be required if

a)   there is a constraint, on the elapsed time allowed for retrieval/storage of a record, which precludes search access (e.g. a real-time system which involves a large file on line), or
b)   the penalties incurred with direct access are not sufficiently severe to outweigh the disadvantages of sequential access.

Part b) of this statement is an extremely complicated issue, involving not only the programs which actually use the files concerned, but also the 'opportunity costs' in other programs which compete for CPU and backing-storage space and time. Such discussion is beyond the scope of this book; the only contribution I wish to make is that if there is no clear advantage of one access method over another, indexed organisation is usually dominant on the grounds of procedural simplicity, support for sequential or direct access in present procedures, potential for eliminating workfiles (especially if alternate indexes are supported) and flexibility to meet unstated future requirements.

For simplicity in the discussion to date, it has been assumed that direct access will be done through an index. This will usually be the case, but the alternative of transforming the primary key to a relative record number is worth considering in special cases. The main disadvantages are:

possible even sparser use of storage than with indexed organisation;
complications if sequential access is required or if it is desired to access all records in a group which share a common partial key;
the need to develop procedures to handle overflow resulting from synonyms.

If there is a large call for retrieval of individual records in a file too large to hold and search in memory, speed of retrieval being important, relative organisation is indicated.

Question

1   Suppose a computer has a content-addressable backing-storage device attached. This device works by quickly making a hardware search of the filestore to retrieve a record of nominated key. Is there a penalty for direct access?   (2 min)

ANSWER POINTERS

Section 11.1

1   I do not know of any objective evidence to support the truth of the proposition to any extent.
If it is true that data analysis helps to uncover loose ends and plan them out of the system, construction does not run so much risk of having an unexpected new obstacle arise. Unexpected new obstacles are very often costly, requiring patching up of the system (often hastily) and reworking of designs or procedures already completed or partially done. Faults in patched-up systems are often more difficult to diagnose – and detection, diagnosis and correction of faults can easily amount to 50% of the effort expended on programming. Patched-up systems are more difficult to maintain when operational.

Final list of files (compare with data model, Figure 7.10):

TRANSACTION (TRANS #, CUST #, TRANS - CLASS, TRANS - DATE, AMOUNT)
(Note: CUST # posted, eliminates TRANS - CUST file.) Sequential, order on trans within cust .
CUSTOMER (CUST #, CUST - CLASS, BREAD - DISC, CONFECTIONERY - DISC, INV - NAME - ADDR, CONS - NAME - ADDR, CURRENT - BALANCE, SIGN - OF - BAL,
          (PREVIOUS - BALANCE, SIGN - OF - PREV - BAL) (3))
Unchanged, indexed organisation.
STANDING ORDER LINE (CUST #, COMM #, REGULAR - REQUIRED - QUANTITY (6), REQUIRED - QUANITITY - THIS - WEEK (6),
                      QUANTITY - DELIVERED - THIS - WEEK (6))
Unchanged, indexed organisation.
The commodity file is split for efficiency since the large amount of sales history data is used only weekly, whereas the 40 characters of commodity details are used
every day for every standing order line .

COMMODITY (COMM #, PACK - QUANTITY, PACK - PRICE, PACK - POINTS, COMMODITY - DESCRIPTION)
Relative organisation, COMM # in range 001 - 499 serves as relative record number.
COMMODITY - SALES (COMM# , (WEEKLY - SALES (8) , MONTHLY - SALES (24)) (24))
Relative organisation.

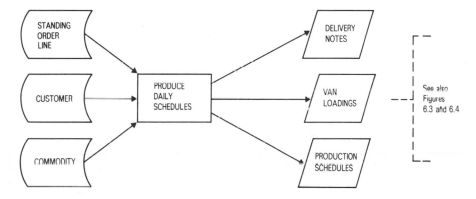

10 900 occurrences of standing order line: will take about 1 minute to read.
300 occurrences of customer: say 10 s.
Each customer has average of 36 commodities. If 499 possible commodities, blocked in 12 s,

the probability of wanting a particular record is 36/499 ;
the probability of not wanting a particular record is 1 - 36/499 ;
the probability of not wanting any record in a particular block is $(1 - 36/499)^{12} = 0.41$

So for each customer 59% of the 42 commodity blocks will be accessed. The commodity file fits on one cylinder, so given an average read
time of 15 1/2 ms, the total commodity file reading time will be 59% × 42 × 300 × 15 1/2 ms = 115 s
Total printing time is estimated:

Delivery lines	10 900
Name, address, total etc. lines, 300 cust @ 10	3 000
Total lines on other reports negligible, say	100
	14 000

Say 14 minutes at 1000 l p m: a printer - bound job.

Fig. 11.2: Example rough calculations for file organisation decisions

If data analysis brings a better **structure** to the design, as I believe it can, this can lead to simpler procedures. Simple procedures are easier to construct, easier to test and easier to maintain. Complexity, to speak rather vaguely, tends to bring an exponential increase in difficulty.

However, it is not certain that such benefits will flow from data analysis in a particular case. At best, though, it reduces the risks that will be run if insufficient planning and insufficient attention to detail are the alternatives.

Section 11.2

1   If the indexes are built on all the desired keys, there is less or no need to resequence the files or extract portions of them for subsequent processing.

Section 11.3

1   Which employees had department X as their previous department? What is the previous department of employee Y?

2   PAYDATA (EMPNO, ...) is passed 9 times per month.
PERSDATA (EMPNO, ...) is passed 8 times per month.
The total characters transferred per employee in a month would be
$$9(x + 150) + 8(x + 100) = 17x + 2150$$
The recombined records would result in characters per month
$$9(x + 250) = 9x + 250$$
Recombination is worthwhile if the recombined transfers are less than the split transfers, i.e. if
$$9x + 2250 \text{ is less than } 17x + 2150$$
$$\text{or } 100 \text{ is less than } 8x$$
$$\text{or } \underline{x \text{ is greater than } 12.5}$$

3   (a) If there are n update and enquiry passes, the split transfers are
$$(n + 1)(157) + n(107) = 264n + 157$$
and the recombined transfers are
$$(n + 1)(257) = 257n + 257$$
Recombination is worthwhile if
$$257n + 257 \text{ is less than } 264n + 157$$
i.e. $100 \text{ is less than } 7n$
$$\text{or } \underline{n \text{ is greater than } 14.3}$$
(b) Let there be n combined passes, j and k passes respectively of the two split files, let the length of the primary key be x and the remaining record lengths be l_j and l_k respectively.
Split transfers:
$$n(2x + l_j + l_k) + j(x + l_j) + k(x + l_k)$$
Combined files:
$$(n + j + k)(x + l_j + l_k)$$
Do not split if combined files transfers are less than split transfers.
In a real-world problem, a further influence on the decisions may be the need to reduce transfers at times when the computer has a peak load of work.

4   (a) $1500(18 + 120) =$ 207 000 characters
   (b) $1500(18 + 3 + 2.6(12)) =$ 78 300 characters
   (c) $1500 \times 2.6(31) =$ 120 900 characters
   (d) $1500 \times 0.55(18 + 1 + 24)$
       $+ 1500 \times 0.40(2(18 + 1 + 24))$
       $+ 1500 \times 0.05(3(18 + 1 + 24)) =$ 96 750 characters
   (e) $1500 \times 0.85(18 + 1 + 36)$
       $+ 1500 \times 0.15(2(18 + 1 + 36)) =$ 94 875 characters
   (f) $1500 \times 0.95(18 + 1 + 48)$
       $+ 1500 \times (0.05(2(18 + 1 + 48)) =$ 105 525 characters

Section 11.4

1   Only in the sense that if such a device is more expensive than other devices, using it may incur a cost penalty. If the device is there anyway, using it is incurring an opportunity cost if other files which are candidates

for the device are displaced to other devices.

REFERENCES

(1) Severance, D. G., and Carlis, J. V., A practical approach to selecting record access paths, **Computing Surveys,** 10, 2, 157–98, December 1977.
(2) Waters, S. J., **Computer system design,** NCC Publications.

12 Computer procedure specifications

12.1 COMPUTER RUN DESIGN

Designing computer runs means specifying the following:

what programs are to exist to operate on the master and transaction files;

what workfiles and security files are needed;

how frequently and in what sequence the programs are to be executed.

The analyst can start by providing a program to update each master file, using the designed inputs. Often, more than one type of update will be conceived for a single master. Simplicity is best preserved by providing a separate program for each type of update. If a customer file contains both name-and-address details and account details, one program may implement name-and-address changes while another may update account details. If the customer file were split for efficiency into a name-and-address file and an account file, such programs would operate independently on their respective files. In this case, if every customer has an account and vice versa, additions and deletions of customers should be carried out in harmony on both sides; a third program may be conceived to do this. Simple programs defined in this way can be tuned later if necessary – see chapter 13.

At this stage, the analyst will have a number of programs of the form shown in Figure 12.1.

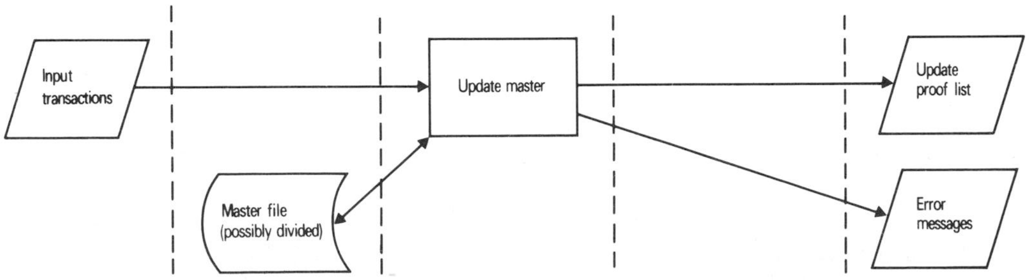

Fig. 12.1: Skeleton of a master file update. This run chart, to National Computing Centre standards, is divided into five columns representing Input Files, Master Files, Processes, Other Files and Output Files

Validation of the input transaction is normally prerequisite to use of it. It is usually convenient to consider validation as a separate program, as in Figure 12.2.

If the master is being updated in transaction mode, the valid transaction 'file' passed from the validation program to the update program will contain only one record at a time. The validation program can in principle carry out the 'database consistency' checks (see section 10.4) by reading the

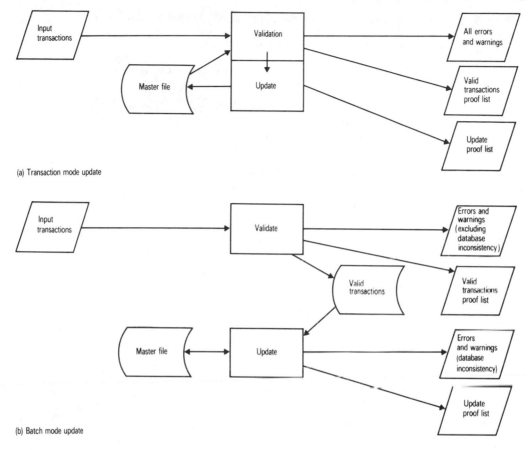

(a) Transaction mode update

(b) Batch mode update

Fig. 12.2: Two possibilities for validation and update

master file. To avoid doubling up on accesses to files on backing-storage, the validation program can pass the valid transaction and corresponding master record, as parameters, to the updating program as in Figure 12.2(a).

If the master is being updated in batch mode, the batch of valid transactions may be passed as a backing-storage file between the validation and update programs, as in Figure 12.2(b). In this case, it is not usually feasible to carry out database consistency checks at validation, and these inconsistencies must be tested for (somewhat unhappily) at the time of update.

In either case, it will be desirable to have a proof list

1   of valid transaction records, so users can be satisfied that all transactions entered have been either rejected or accepted correctly, and
2   of master file records as updated by the transactions, so users can be satisfied that the master file has been correctly updated by those transactions.

If such lists are not to be used as everyday or historic records of the system, but only for initial proving and subsequent spot checks, there should be some means (e.g. operating system parameter) of suppressing the production of the proof lists when they are not wanted, or perhaps writing them only to backing storage for subsequent printing should the need arise. Often, it is decided to suppress update proof lists apart from control totals, which are checked every time. Transaction proof lists and batch totals are usually destined to become a permanent hard copy record of all transactions entered. These and other integrity and security aspects are discussed in chapter 14.

If the master file is to be updated by sequential or skip-sequential access of the records, then the valid transaction file will need resequencing into the order of the master file (this assumes that the transaction file is to be accessed sequentially and arises unordered, of course). It will often be profitable to implement a sequential-access update by reading in the present copy of a master and writing out a fresh updated copy, as shown in Figure 12.3. A by-product of this approach is that a security copy of the file exists in the form of the old master and the transaction file, which can be used for re-creation should the new master be destroyed or corrupted. If the master had indexed organisation, a further by-product is that it is reorganised.

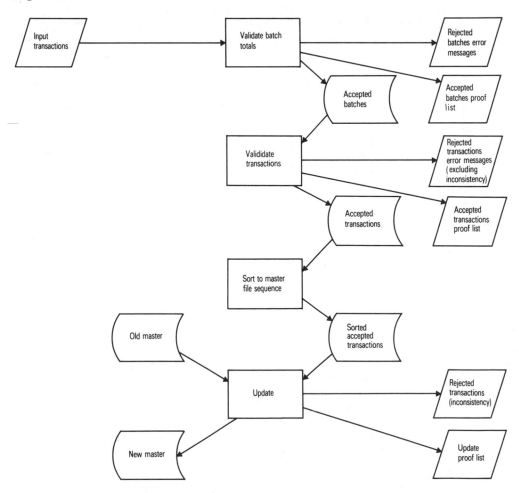

Fig. 12.3: Illustrating a batch update of a master on a brought-forward-file, carried-forward-file basis. Also illustrating one way in which a further workfile, and splitting the validation program, can overcome limited printer facilities if separate reports about accepted/rejected batches and accepted/rejected transactions are required

In the examples used to illustrate these concepts, it is assumed that rejected transactions in a batch can be delayed until the next processing period, so the accepted transactions can be put forward to update the master without waiting for the corrections. This is not always the case. With a payroll program, for example, it is usually considered that all rejected transactions should be resubmitted before running the payroll. In this case, if the update program cannot be run repeatedly, the validation program will need to be able to add the resubmitted transactions to the file of previously accepted transactions.

Now the analyst can provide a program for each other required output. Again, simplicity is preserved by considering a separate program for each report. If the report is in a sequence different from that of the file used to create it, and the report is too large to be resequenced in memory, either an alternate index or a sorted workfile will be needed. Usually, these reporting programs use the files only as input, although sometimes the records are updated, e.g. a status flag is set.

It is quite a common requirement for reports to compare a current status of files with a previous status, e.g. a summary sales analysis report comparing this month's sales with those of the same month last year. Usually the most practical way of handling this requirement is to store in a workfile the summary figures relating to the previous period. This workfile is conveniently created while preparing the report in the previous period. The content of the workfile is simply the relevant content of the report being created. Figure 12.4 illustrates a typical case.

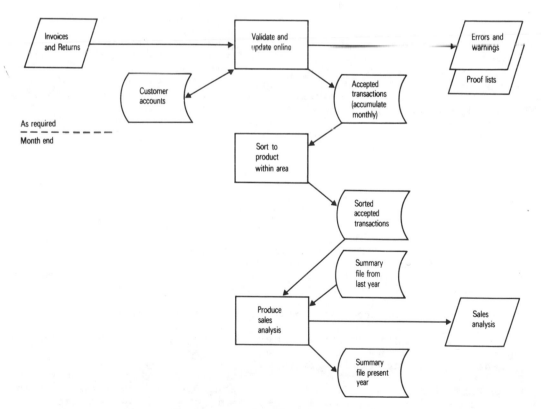

Fig. 12.4: Outline of a system to produce comparative sales statistics

Questions

1   With the system illustrated in Figure 12.4, the sales analysis reports produced in the first year of the system's operation bring a problem. What is the problem and how can it be overcome?   (5 min)

2   Draw a computer run chart for a daily validation and sequential update in which all rejected transactions are logged on magnetic tape, with uncorrected transactions being reported monthly, in age sequence. All the details of rejected transactions are to go on the log tape, and there is an indefinite number of rejections possible. Input transactions are unsequenced.
                                                                (30 min)

3   What options are open if a monthly report is to show the current month's transactions summarised, the last three months transactions and the last twelve months transactions?   (10 min)

12.2 COMPUTER PROCEDURE SPECIFICATIONS

A large number of techniques for specifying programs have been proposed and widely adopted. Most of them have at heart a structured approach and some graphic aid. Some go as far as a pseudo-code or program definition language with well-defined grammatical structure and rules. A popular approach where decision table preprocessors are extensively used is to specify procedures exclusively by decision tables.

This is another of those cases where there are many ways to succeed. The method I prefer is to describe the procedure algorithm in Structured English, without a graphic aid. This method is particularly easy to modify and maintain.

The rules are very simple. The analyst describes the algorithm of the program in natural language, but he limits his control structures to those compatible with structured programming – in other words, nc GO TOs – (or at any rate, no backward GO TOs). This means that specifications will take the form of sequences of natural-language imperatives, possibly embedded in IF...THEN...ELSE... constructions or CASE constructions, under the control of expressions such as WHILE 'condition' ...; UNTIL 'condition' ...; for EACH record ...; REPEAT n TIMES ..., etc. The roguish statement WHENEVER condition such-and-such arises, do so-and-so, is perfectly legitimate. The analyst can choose any control expression or imperative that preserves structure, precision of meaning and clarity of expression. Since the aim is to communicate the algorithm to a programmer, preference is naturally given to those phrases that occur in the programming languages used locally, or which fit in with local vernacular and group norms.

The scope of a control loop or condition is shown by indenting the subordinate statements by a few spaces. At the end of the controlled or conditional statements, the indentation is dropped (use paper with vertical rules, such as the NCC chart sheet). Thus a statement starting in a margin is the one that applies when the previous control statement in that margin is exhausted. An ELSE is always lined up with the IF or other conditional expression which it matches. This simple procedure is a very powerful visual clue, and there is no need for strange keywords like ENDIF or ENDDO.

Possible ambiguity of control statements (e.g. UNTIL condition ...; is the condition to be tested before or after the first cycle?) is unimportant if context, usual practice and the programmer's common-sense can be relied upon to make the right choice. Similarly, there is no need to spell out every detail and initial condition when the programmer can be relied upon to give a correct interpretation. It follows that the level of detail given in the specification will be dictated by the amount of knowledge and initiative expected from the programmers in the particular working environment.

PAYROLL PROGRAM

<u>Print initial messages</u>
For each employee
   |<u>Calculate hours to gross</u>
   |<u>Calculate gross to nett</u>
   |<u>Produce a payslip</u>
   |<u>Produce a credit transfer</u>
   |<u>Print a payroll line</u>
   |<u>Accumulate payroll totals</u>
<u>Print payroll totals</u>
<u>Record payroll trailer</u>

(Underlined statements are expanded in subroutines, not illustrated)

CUSTOMER NAME AND ADDRESS UPDATE PROGRAM

Print report heading
For each name-and-address change
   |Retrieve name-and-address master with corresponding customer number
   |If invalid number
   | |customer is not present
   |Else
   | |customer is present
   |If amending or deleting change
   | |If customer not present
   | | |Print error message A
   | | |Add 1 to error count
   | |Else
   | | |If deleting change
   | | | |Delete master record
   | | | |Add 1 to deletions count
   | | |Else (amending change)
   | | | |<u>Update master record</u>
   | | | |Add 1 to update count
   |Else (inserting change)
   | |If customer is present
   | | |Print error message B
   | | |Add 1 to error count
   | |Else
   | | |Insert new master
   | | |Add 1 to insertions count
Whenever linecount = 55 above
   |Print page footing
   |Increment page count
   |Print page heading
Print report footing

(Assuming adequate accompanying print and record layouts, maybe only 'update master record' needs amplification as a subroutine)

Fig. 12.5: The program definition language, Structured English. Arbitrary subroutines are created, as desired, by the use of the underlined phrases. Brackets may enclose comments, subscripts, arithmetic expressions, etc. as desired: the meaning is clear from context. It is recommended that when an ELSE is more than once removed from its IF, the converse of the condition be shown as a comment

NIM GAME PROGRAM

(Players remove up to three counters in turn. The player removing the last counter loses)

```
Display "THE NIM GAME. THERE ARE 14 COUNTERS"
While a game is required
    |Set no. of counters to 14
    |Set count to zero
    |If coin equals heads
    |    |make it user's go
    |Else
    |    |make it computer's go
    |Until no. of counters is less than 2
    |    |Play the game
    |If no. of counters = 1 and it is computer's go
    |    |Print "YOU WIN"
    |    |Add 1 to user's score
    |Else
    |    |Print "I WIN"
    |    |Add 1 to computer's score
    |Print "DO YOU WANT ANOTHER GAME? REPLY YES OR NO"
    |If answer is "NO"
    |    |a game is not required
Display "I WON" computer's score "GAMES"
Display "YOU WON" user's score "GAMES"
Stop
```

Coin equals heads subroutine

```
Get time from operating system
If even number of seconds
    |coin equals heads
```

Play the game subroutine

```
If user's go
    |Set N to zero (an invalid value)
    |Until N is valid repeat
    |    |Print "YOUR GO - TAKE 1, 2 OR 3 COUNTERS"
    |    |Read N
    |    |If N is not in range 1-3 inclusive
    |    |    |Print "NO CHEATING"
    |    |    |N is not valid
    |    |Else
    |    |    |N is valid
    |Make it computer's go
Else (computer's go)
    |Case of count
    |    |0: set N to 1
    |    |1: set N to 1
    |    |2: set N to 3
    |    |3: set N to 2
    |Print "MY GO - I TAKE" N "COUNTERS"
    |Make it user's go
Take N from no. of counters
Print "THERE ARE" no. of counters "COUNTERS LEFT"
Set count to (count + N) modulo 4
```

Fig. 12.5 continued

MAIN PROGRAM

Display "THE NIM GAME. THERE ARE 14 COUNTERS"

While a game is required

Set number of counters to 14

Set count to zero

Coin equals heads? Y / N

Make it user's go | Make it computer's go

Until number of counters is less than 2

Play the game

No. of counters = 1 and computer's go? Y / N

Print "YOU WIN"
Add 1 to user's score

Print "I WIN"
Add 1 to
computer's score

Print "DO YOU WANT ANOTHER GAME? REPLY YES OR NO."

Answer — "NO"? Y / N

A game is not required |

Display "I WON" computer score "GAMES"

Display "YOU WON" user score "GAMES"

Stop

COIN EQUALS HEADS

Get time from operating system

Even numbers of seconds? Y / N

Coin = heads | Coin = tails

PLAY THE GAME

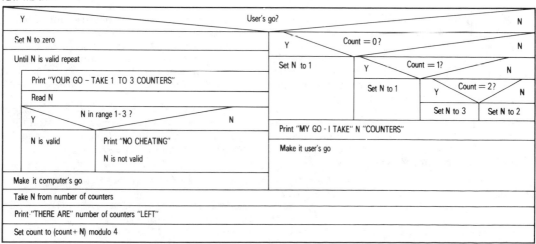

User's go? Y / N

Y side:
Set N to zero

Until N is valid repeat

Print "YOUR GO – TAKE 1 TO 3 COUNTERS"

Read N

N in range 1-3 ? Y / N

N is valid | Print "NO CHEATING"
N is not valid

Make it computer's go

N side:
Count = 0? Y / N

Set N to 1 | Count = 1? Y / N

Set N to 1 | Count = 2? Y / N

Set N to 3 | Set N to 2

Print "MY GO - I TAKE" N "COUNTERS"

Make it user's go

Take N from number of counters

Print "THERE ARE" number of counters "LEFT"

Set count to (count + N) modulo 4

Fig. 12.6: Nassi–Shneiderman diagram of the NIM game program

The imperatives chosen can be at as high a level as may be desired. They can be amplified, in the manner of subroutines, if necessary. If you are not familiar with structured programming, you may find that quite a lot of practice is needed to acquire skill in Structured English. You would be best advised to practise with Nassi-Shneiderman diagrams first. Figure 12.5 shows examples of procedure specifications written in Structured English and Figure 12.6 takes the last example to illustrate the relationship between Structured English and Nassi-Shneiderman diagrams.

If you have a background in programming, you should try to put this aside and concentrate on a fine use of words to describe algorithms, independently of programming techniques. There is no need to solve programming problems in the program specifications - providing, of course, the analyst is confident they are soluble. The specification should concentrate on the business procedures and should be written keeping this objective in view without being side-tracked into solving programming problems. For example, there is no need to 'set flags' in Structured English (barring a flag field in the data). A 'flag' at this level should signal a condition which is visible in the business problem and there must be an English phrase to cover the point. Any programmer worth his salt will be able to translate this into a computer procedure. For another example, suppose punched cards are being input with three transactions per card. Programmer's thinking often starts the algorithm 'For each card ...'; analyst's thinking should start the algorithm 'For each transaction ...'.

Question

1   Assume that a file exists containing invoice records as follows:

salesman number; salesman name; invoice number; invoice amount.

There is an indefinite, but large, number of invoices representing all those issued over a twelve-month period. Outline a computer procedure specification to produce from this file a contribution-by-value report as illustrated in Figure 3.6. There is no definite maximum number of salesmen.   (20 min)

12.3 DOCUMENTING THE PROGRAM SPECIFICATION

A complete program specification is the collection of documents handed to the programmer as his brief. The original documents will exist in the system files (see section 6.1), so only copies are handed to the programmer. The extent of the documents and the level of detail in them depends on local conditions. The greater the division of function between analysts and programmers, the more the program specification must be complete, self-contained and explicit.

The following is a reasonable target content of program specifications passed between internal analysts and programmers working in a functionally-divided Data Processing department. It assumes that the programming section has the responsibility to further divide the programs down to the module level where necessary.

1   **System Outline** at a level to summarise the program's inputs, outputs, processes and files (see section 6.2).
2   **Computer Run Chart** showing the context of the program - where its inputs derive from, where its outputs are destined (see section 12.1).
Alternatively, a **Program Procedure Summary** (see NCC standard 3.12 - not illustrated here) can be used instead of both the above.
3   **Computer procedure specification** in Structured English (see section 12.2) or other local standard, supported or supplanted by computer procedure flowcharts, interactive system flowcharts (see section 8.5) and decision tables (see section 6.3), as appropriate.
4   **Computer File Specification** for each file (apart from printer, display and optically-read files) and Record Specifications for each record. See NCC standards 4.1.2 and 4.2.1.

5   **Computer Document Specification** (NCC standard 4.1.3) and **Print Layout Chart** for each document or printed listing, **Display Layout** for other displays (see section 9.3). Printed layouts or displays may be amplified on Record Specifications in complicated cases – consult NCC standards for this usage.

6   **A statement** of testing intentions and other programming objectives. This is not to say what test data is required, but to define responsibilities for testing so that the programmer knows how far he should go in testing the links between this program and the ones preceding or following it, whether he should test the interactive procedures with the users, etc.

Questions

1   How might the specification content be reduced if the analysts and programmers work together in small teams, or analyst/programmers are employed?   (3 min)

2   What could be covered in 'Other programming objectives'?   (5 min)

3   What should be supplied if it is the analyst's task to define program structure and modularisation?   (3 min)

ANSWER POINTERS

Section 12.1

1   There is no summary file from last year. **Either** the sales analyses in the first year will have to be produced without the previous year's comparative sales (in which case the program specification must allow for the optional input file) **or** the sales statistics must be 'converted' from the present system to create an initial summary file for last year. Section 16.1 deals with conversion.

2   See Figure 12.7 overleaf.

3   One possibility would be to follow the philosophy of Figure 12.4 and produce a summary workfile of each month's sales. This could get a little complicated.

Another possibility is to maintain a file in which each report category has 12 fields, one for each of the last twelve months. At month end, the totals for the oldest month are dropped ('rolled out') and the totals for the most recent month added ('rolled in'). These rolling totals can be used to satisfy the reporting requirements.

Section 12.2

```
1   Resequence the invoice file to salesman number order
    For each salesman
        |For each invoice record
        |    |Accumulate salesman total
        |    |Accumulate file total
        |Write salesman number and salesman total to workfile
    Resequence workfile to salesman total order (descending)
    Write report headings
    Set rank, cumulative to zero
    For each salesman in workfile
        |Increment rank by 1
        |Calculate % total = (salesman total ÷ file total) x 100
        |Add salesman total to cumulative
        |Write report detail line
```

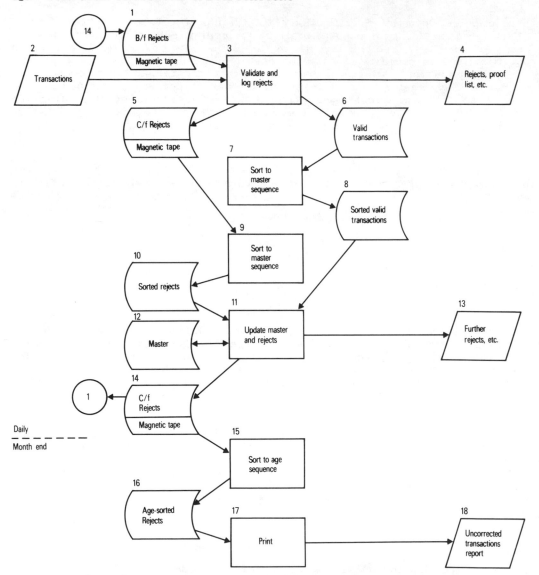

Fig. 12.7: Answer pointer to question 2, section 12.1. It is assumed that the additional rejects are added at the end of the rejects file in step 3; failing this, a better solution may be to merge file 1 into the sort at step 9

Section 12.3

1   If analysts and programmers are completely integrated, only the statement of objectives is 'new' – the rest of the documentation is in the files and the programmer knows how to get it. So in the **least** case, the 'brief' need only point to the system documentation – a System Outline could do this.

2   A guide to the programmer as to the priorities and targets of this particular program – deadline for delivery, need for special efficiency or quick response, need to keep the program small, need for portability, etc.

3   A module (component) plan and a specification for each module.

REFERENCES

(1) Nassi, I. and Shneiderman, B., Flowchart techniques for structured programming, **SIGPLAN Notices, 8,** 8, 12–26, 1976.
(2) Stay, J. F., HIPO and integrated program design, **IBM Systems Journal, 15,** 2, 143–54, 1976.
(3) Van Leer, P., Top–down development using a program design language, **IBM Systems Journal,** 15, 2, 155–69, 1976.
(4) Program design techniques, **EDP Analyzer,** 17, 3, March 1979.

13 Efficiency of computer procedures

13.1 TRANSPARENT TUNING

Chapters 11 and 12 have offered a method of computer system design which emphasises finding simple, working solutions rather than technically efficient designs. This chapter suggests some methods for tuning up the efficiency of a system.

The first question to be considered is whether or not this tuning is needed. Maybe the analyst and programmer resources would be better devoted to the next application. Maybe more hardware will cheaply remove the need for tuning. Maybe a review of operating procedures, downtime, scheduled maintenance and non-operational uses of the computer, such as program testing, will reveal scope for increasing the supply of computing resources. Maybe the tuning will not result in saving a shift, overtime or machine upgrades, nor provide an improvement in service to the users. In this case, the value of the increase in the amount of slack computing resource is uncertain. It may be worthwhile to defer tuning until it is more urgently required, particularly while hardware prices continue to fall.

Tuning usually has an objective of meeting a constraint of the computer power, or maximising throughput by minimising contention for limited computer resources (memory, CPU time, input–output time, backing storage), or satisfying, or improving, response time and deadlines (minimising elapsed time). Some tuning actions can improve throughput **and** elapsed time. Others may improve one at the expense of the other. The discussion that follows emphasises maximising throughput by minimising computer resources used in file accesses and transfers, rather than by minimising CPU time or space requirements. (This is a simplification, because the two are interlinked. Tuning procedure code to reduce space requirements in memory leaves more space for buffers. Tuning code for reduced CPU time often increases the space needed for procedures. A reduction in data transfer, with some computers, can reduce CPU cycle stealing proportionately.)

The computer runs in the design that are expected to use the most resources, or take the longest time, should be identified and considered as the first candidates for tuning. Often, these will be the runs that involve sequential passing of a large master file and, of course, any run that is made frequently. The point is that there is no cause for spending time in tuning programs that use only a trivial amount of resource.

Considering in turn each of the programs needing to be tuned, the 'bottle-necks' in the program execution should be identified. These are the resource usages which are expected to hold up completion of execution. Often, in commercial applications, these will prove to be the input–output resources such as device connect time, channel connect time, seek and search time. If both input–output and CPU time are considered bottlenecks, tuning should first concentrate on increasing simultaneity by spooling, buffering or operating system enhancements.

Having identified the important bottlenecks, consideration should be given to the following.

1 **Transparent tuning** This is tuning done through the operating system, not involving recompilation of programs. Transparent tuning does not

increase the complexity of the system.

2   **Superficial tuning**   This affects only the detail of the programs' use of files and does not require revision of the computer runs nor the basic algorithms of the programs.

3   **Hardcore tuning**   This requires revision of the algorithms or changes to the runs through merging or separation of programs and files. The file splits mentioned in chapter 11 were hardcore tuning decisions easily made at the outset and easy to implement providing the decisions are made early. With further hardcore tuning there is usually, but not always, quite a large penalty in increased complexity, as measured by the difficulty experienced by the programmers when they come to understand and implement the design and maintain the system after implementation. There may also be a penalty of inflexibility in adapting the system to future needs. A design in which the imprint of the data model can be clearly seen is generally flexible. If the data about entities and relationships is widely scattered or duplicated in different files, or mixed in one file, revisions may pose difficulties.

These three tuning decisions are listed above in order of simplicity but, unfortunately, in probable reverse order of size of the improvements in efficiency they bring. Often it will be clear from the outset that hardcore tuning must be done if cost or equipment constraints are to be met.

What tuning actions are transparent in a particular case depends on the facilities of the operating system. Which actions are good actions is sensitive to the characteristics of the hardware and software concerned. The following are likely possibilities.

a   **Device allocation**   Allocate a bottleneck file to the fastest suitable device.

b   **Buffering**   In a large computer, where many jobs reside at a time, the operating system usually tries to maximise simultaneity. If this is the case, double buffering at the program level is probably counter-productive; it simply increases CPU cycles and memory used for input-output, with no increase in overall simultaneity. The memory tied up may increase overlay and paging requirements, leading to degraded overall performance. In such cases, single buffering should be the rule. In other cases, double buffering should be used (assuming this does increase simultaneity with the device concerned, which is not always the case). In rare cases, where a program is alternately CPU bound and input-output bound, triple or more buffering may be advantageous.

c   **Blocking**   For sequentially organised files, the optimum block size, B_i of file i, from the viewpoint of a single program simultaneously searching files 1, 2, ..., i, ..., n, is (after Waters, 1):

$$B_i = \frac{S\sqrt{D_i G_i I_i}}{D_i \sum \sqrt{D_j G_j I_j}}, \quad j = 1, 2, ..., n$$

where S is the total space for buffers in units such as characters, D_i is the number of buffers allocated to the file, G_i is the inter-block time penalty, I_i is the total file size in units.

Since several programs may use a file, the blocking factor should ideally minimise total input-output time for all programs. In practice, there is only a small gain from further optimisation once large block sizes have been allocated to large files. This is because input-output time is rather insensitive to block size when large blocks are involved. The analyst might as well independently consider the programs using the files, and strike an arbitrary balance.

Estimating S before the programs are written is not easy. In the case of large, multi-task computers, S should be estimated bearing in mind that too generous a view of space available may lead to degraded overall performance through increased overlay and paging.

Questions

1   A sequential transaction file updates a sequential master by copying forward. There are 100 000 masters updated by 10 000 transactions. The master record is 250 words long and the transaction record 100 words. It has been decided to double buffer all files. The inter-block gap time is 5 ms. 8000 words of storage are available for buffers. What block sizes would you recommend?   (20 min)

2   List six other possible tuning attributes of files which may be changed transparently, given a comprehensive operating system.   (5 min)

13.2 SUPERFICIAL TUNING

Taking file organisation and access as given, the principal strategy for superficial tuning is to reduce the number of bits used to store the records on backing storage, i.e. 'data compression'. Record layout (Figure 11.1) is assumed to be decided hereafter, although the original decision, which struck a reasonable balance between file size and system complexity, might be revised if further tuning proves necessary. Techniques 4, 6 and 7 below call for variable-length records.
 Data compression techniques include the following.

1   For all-numeric data, use of implied decimal points and use of packed decimal representation (4-bit code). Sometimes a further small saving is possible by using binary fields, especially if they can be unsynchronised to word boundaries.
2   For alphanumeric data, use of BCD (6-bit) codes instead of EBCDIC (8-bit) if 64 codes are enough (e.g. no lower case data).
3   For alphabetic data, use of 5-bit codes if 32 codes are enough.
4   Use of variable-length fields to eliminate leading/trailing blanks, zeros. Such fields can be preceded by a length count or can be given an end-of-field marker (a character not otherwise appearing in the data).
5   For a small (less than 1000?) set of relatively stable symbol sequences (e.g. names of countries, departments, weekdays) which are often repeated in the data, assign a substitute brief code and use a look-up table to translate. Should the symbol sequence be embedded in a variable-length value, the brief code will have to be identified by (or consist of) a symbol not otherwise appearing in the data.
6   If some symbols (e.g. blanks, zeros) appear often in runs of 2 or more, consider substituting unused characters for a pair, or, if runs of three or more are common, an unused character followed by a single digit which represents the length of the run in excess of 3 (e.g. 3 blanks represented by ¥0, 5 blanks by ¥2).
\7   Consider coding based on frequency of occurrence. From an information-theory point of view, the use of fixed-length codes (e.g. BCD, EBCDIC) is wasteful of bits if the symbols do not have equal probability of occurrence. It is possible to assign unequal length codes – short codes to common symbols, long codes to rare symbols. For example, taking a trivial system where only three symbols are used:

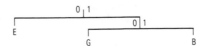

Fig. 13.1: Asymmetric code tree for a three-symbol system

EGG would be coded 01010; BEG, 11010; BEE, 1100.

To construct an efficient tree, list the symbols in order of frequency of occurrence. Divide the list into two parts at the point which most nearly gives equal total frequencies in both parts. Assign a first bit of 0 to symbols in one part and a first bit of 1 to symbols in the other part. If any part contains more than one symbol, bisect that part again at the mid-frequency point into two further parts (assigning a zero for the second bit of the symbols in one of the further parts and a 1 for the second bit of the symbols in the other further part; similarly for third, fourth etc. bits as required). Continue until all parts contain only one symbol.

It is not reasonable to generalise about the complexity of these alternative methods. For example, some machines have operating systems which permit BCD code to be selected transparently, while at the other extreme some machine/device combinations require assembler language routines to translate EBCDIC to BCD and to arrange for the BCD-coded symbols to be adjacently stored (transgressing byte boundaries). The analyst can weigh up the consequences of data compression on complexity only in the context of a particular computer.

Question

1   Suppose it is decided in an installation to have a standard subprogram, used by all programs referencing backing storage, which will compress data records just prior to being stored (all stored records will be of variable length), and decompress them immediately after they are read (so that application programs are unaffected by compression techniques). The files have upper and lower case alphanumeric data, with at least 3 punctuation symbols in use.
(a) What compression techniques would you consider for (i) the subprogram, (ii) the application programs?
(b) If the subprogram uses asymmetric encoding calculated from analysis of the actual frequency with which EBCDIC codes appear in the data, is there any advantage in using any other compression technique either in the subprogram or in the application programs?   (25 min)

13.3 HARDCORE TUNING

The main strategy for hardcore tuning to reduce input-output time is to combine runs so as to eliminate workfiles or repetitive passing of files. The penalties to pay are usually increased program space requirements or increased overlay or paging activity, increased programming and testing complexity and reduced ease of maintenance or adaptation to new require-ments.
A simple example of run combination can be illustrated using the system shown in Figure 12.4. The Sorted Accepted Transactions File, let us suppose, is not required for any purpose other than as input to the Produce Sales Analysis program. By merging the Sort program into the Produce Sales Analysis program so that the transactions returned in the last pass of the sort are handed directly to the sales analysis routine, the Sorted Accepted Transactions File is eliminated (Figure 13.2).
The increased complexity in this example is small or nil if 'own coding' on the last pass of the sort is supported in the high level language in use locally (e.g. COBOL SORT verb with OUTPUT PROCEDURE), but it might be a more substantial obstacle otherwise.
Figure 13.3(a) illustrates a master file being updated by two different types of transaction. The double pass of the master file can be eliminated by having one program whose job is to pass the master, calling upon the other update programs to match their transaction files and update the master record passed to them as in Figure 13.3(b). This ruse increases programming

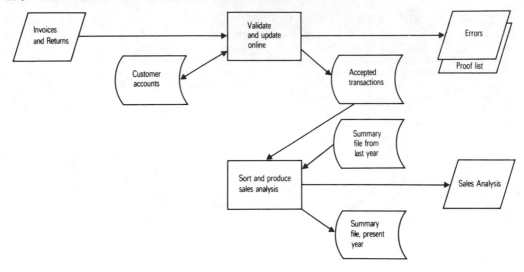

Fig. 13.2: **A workfile eliminated from Figure 12.4**

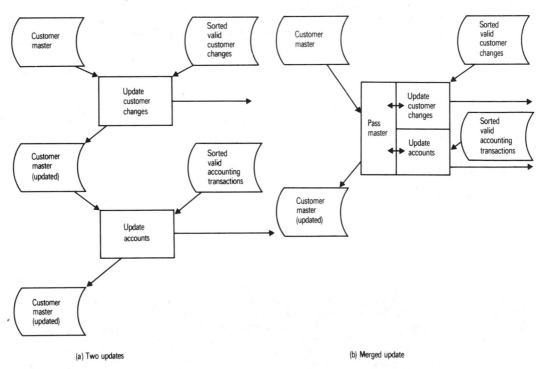

(a) Two updates

(b) Merged update

Fig. 13.3: **Eliminating a master file pass**

and testing complexity somewhat, but otherwise changes the design concept little. A more fundamental approach would combine the two transaction files into one, either by merging them after they are created from the validation programs (Figure 13.4(a)) or by intermixing the original transactions and offering them to a combined validation program (Figure 13.4(b)). A possible compromise would be to have two validation runs as in Figure 13.4(a) but

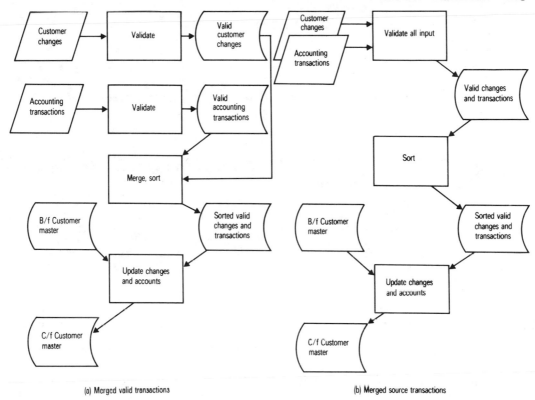

Fig. 13.4: Variations on an update tuned by merging programs. Designing the system like this in the first instance can lead to great complexity. A good discipline is first to develop a system with single–record–type files/ single–function programs, then to wed the components together

to add the output of the second run to the end of the file of valid transactions produced by the first run; the combined file could then be input into the sort as in Figure 13.4(b).

Questions

1   From Figure 13.2, how could the file of 'accepted transactions' be eliminated? What consequences would this bring?   (5 min)

2   From Figure 13.2, eliminating the workfile has saved (a) writing the workfile out from the sort program and (b) reading the workfile into the sales analysis program. Considering the computer as a single–programming machine, can both of these be counted as saving total elapsed time? (5 min)

3   From Figure 13.4(a), draw the run diagram that would result from eliminating the sort/merge by calling the two validation programs as input procedures on the first pass of a multiple–input–file sort, and the update program on the last pass.   (10 min)

4   If a job is printer–bound, what options are open to speed it up? (5 min)

ANSWER POINTERS

Section 13.1

1   Input master $\sqrt{\text{DGI}} = \sqrt{2 \times 5 \times 25 \times 10^6} = 16 \times 10^3$

$$\text{Output master} \quad \sqrt{DGI} = \sqrt{2 \times 5 \times 25 \times 10^6} = 16 \times 10^3$$
$$\text{Transactions} \quad \sqrt{DGI} = \sqrt{2 \times 5 \times 1 \times 10^6} = 3 \times 10^3$$
$$\Sigma\sqrt{DGI} = 35 \times 10^3$$

$$\text{Input master blocksize} = \frac{8000 \times 16 \times 10^3}{2 \times 35 \times 10^3} = 1800 \text{ approx., say } \underline{1750}$$

$$\text{Output master blocksize} = \frac{8000 \times 16 \times 10^3}{2 \times 35 \times 10^3} = 1800 \text{ approx., say } \underline{1750}$$

$$\text{Transaction blocksize} = \frac{8000 \times 3 \times 10^3}{2 \times 35 \times 10^3} = 350 \text{ approx., say } \underline{500}$$

Blocksize has been rounded to a multiple of the fixed length record size in the case of the masters; this rounding down releases space, which has been allocated to the transaction file.

2   Channel assignment, areasize (incremental space allocated to a file when allocated space is exhausted), whether crunching is done or not (crunching releases unused allocated space at file close), file geography (physical location on device, adjacency to other files), recording density, print density, printing speed, transmission speed, character code, indexed file attributes (keys per index entry, number of levels, first level overflow size, etc.).

Section 13.2

1   (a) (i) At least 7 bits per character are needed (if symmetric codes) so the subprogram couldn't just convert EBCDIC to BCD. And it can't pack decimal characters since a pair of packed decimal numbers make an EBCDIC code. Possibilities:

a)   convert 8-bit code to 7-bit code if total characters used are 132 or less;
b)   substitute unused characters for any known regular patterns (e.g. for two spaces);
c)   use run-length coding;
d)   use statistical coding (asymmetric codes).

Note that all the first three rely upon the application programs observing a convention of supplying character data (or data drawn from a code set with restricted possibilities). With binary data, one cannot rely upon any 'character' being unused.
(ii) If the subprogram depends upon a limited set of characters being presented for compression, only coding through a look-up table or, possibly, run-length coding is possible. If special identifying symbols are needed, these must be drawn from the set of symbols that the subprogram considers 'used' by the application programs as a whole, but which is 'unused' as far as the particular program is concerned. (Packed decimal is also a possibility if the subprogram considers the used characters to include all possible packed pairs of digits.)
If statistical encoding is being used in the subprogram, any compression technique can be used in the application programs. The optimal asymmetric code should be derived using the frequency of occurrence of symbols after compression has been done in the application programs.
(b) Yes, probably. The asymmetric code assumes that the probability of a given symbol occurring is independent of the occurrence of another symbol. In practice, symbols describing data in commercial files are auto-correlated, i.e. they have pattern. The other compression methods tend to reduce the pattern in the symbols presented to the subprogram.

Section 13.3

1   Combine the validation/update program with the sort by calling it as an input procedure from the first pass of the sort. Consequence: probable

reduction of online service; increased complexity; no backing-storage record of individual accepted transactions. Altogether probably not a good idea!

2   Only if both would have added to the elapsed time in the first place. This is probably the case with the file created by the sort, but with devices, controllers and channels maybe the reading of the input workfile into the sales analysis program would have been overlapped with the other file processing and printing.

3

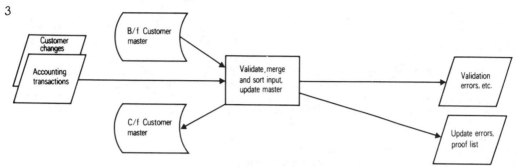

Fig. 13.5: Answer pointer to question 3, section 13.3

4   Use a faster printer, reduce form depth, print two up, increase print speed, reduce number of lines, review special character usage if printer sensitive to these.

REFERENCE

(1) Waters, S. J., Blocking sequentially processed magnetic tape files, **Computer Journal**, 14, 2, 109–12, May 1971.

14 Integrity and security of the data

14.1 PREVENTION

The integrity and security of the data are considered here from the point of view of freedom from risks. The 'risks' are those events that threaten the organisation's data: threaten to destroy or corrupt it or prevent its use, threaten to access it illicitly or to steal it, threaten to use it for illicit purposes. The risks may be of physical loss or damage, corruption by faulty machinery or environment, erroneous or inconsistent data produced by human or program error, malicious damage, deliberate change to data or procedures for purposes of fraud or embezzlement or espionage, theft of data, interruption to processing by machine failure or power failure or communications failure, strikes and so on.

  The designer, as far as he is able, should analyse all the threats posed to a particular system and weigh up the severity of the consequences. Some systems, e.g. batch accounting systems, may be intolerant of error but quite tolerant of delay in processing. Others may be less tolerant of delay but perhaps more tolerant of error. The priorities of the particular case need to be decided and in this light procedures instituted for the following integrity and security measures.

a) Prevention: minimising the chance of the threatened event happening, and minimising the consequences when it does.
b) Detection and proof: discovering that data is lost or corrupted, or procedures improperly executed, or proving that they were not.
c) Recovery: restoring the data and procedures to their proper state.

  In addition to those described in section 10.4, data processing department personnel should consider the following prevention measures.

1) Secure site for computer room and file storage areas. Suitable physical precautions against fire, water damage, etc. Consider standby hardware, generators, duplicated systems. Use reliable suppliers.
2) Provision of fallback procedures in the event of prolonged failure. Provision of emergency procedures should all required data not be available by the system deadlines.
3) Only operations personnel permitted in the computer room or data storage areas.
4) Independent visual check of source code written, and procedures to ensure that object code in production libraries comes from the approved source code.
5) A log, preferably kept automatically by the operating system, of all runs made which access production files, and a log of all updates to the production program library. Authorisation of senior personnel required for amendments to production runs, changes to production schedules and operation instructions, and execution of library updates.
6) A log, preferably kept automatically by the operating system, of all files produced; suitable visual labelling on removable storage; suitable computer checking of file labels to ensure that the correct file/version is being used and to ensure non-overwrite on files whose retention period has not expired.

7)   Participation of internal and external auditors.

Questions

1   What type of participation (see section 1.3) would you suggest for internal and external auditors?   (5 min)

2   If the computer operating system checks that the correct file/version required by a program is supplied to it, does this guarantee that the program runs with the right files?   (3 min)

14.2 DETECTION AND PROOF.

For the purpose of this explanation, it is assumed that the standard error-detection hardware and software supplied by the computer manufacturer is such that if a program reaches normal termination, the reading and writing of data to and from the files may be assumed to have been faithfully done in accordance with the program's instructions. (If the analyst is considering using untried, unfamiliar machinery, this assumption should be questioned.)
  Detection is considered from the point of view of the program terminating abnormally or normally.

1   Program terminates abnormally   'Abnormally' in this context means that the program does not reach any of the end points which at face value indicate a successful execution.
a   The program may have reached another end point, contemplated by the programmer, which is invalid because of some internal inconsistency. This indicates either a fault in the program or, more likely, a fault in an earlier program which left the files in a corrupt or inconsistent state. Local standards should require programmers to arrange for a clear message to be given to the operators in these cases. The message should also identify the source line at which termination occurs, and should give other diagnostic information, such as number of records read and written before the error, if these are not supplied by the operating system.
b   The other possibility is that the program terminates abnormally at a point not contemplated by the programmer. This will arise either because of some external interruption which terminates the program (e.g. power failure, operator intervention to cancel the program) or a fault has been detected and the operator or operating system has terminated the program as a result (e.g. uncorrectable read/write error, program fault resulting in illegal memory reference trapped by operating system).
  These possibilities are serious in a production system and re-emphasise the need for thorough program and system testing (chapter 16). In the present context, though, they are the happiest possibilities since the error is plain to see. The problems lie in diagnosing and correcting the program faults and restoring the correct database status (section 14.3).

2   Program terminates normally   The more perplexing possibility is that the program reaches normal termination when in fact an error has occurred, but has gone undetected.
a   The data may have been incorrectly recorded at origination. There is nothing the computer can do to detect errors in data which was incorrectly, but plausibly, recorded at source. About the only hope here is the auditor's technique of taking a sample transaction (such as an invoice) as recorded on the computer and going back behind the original transaction to verify the details (by asking the customer to verify that the invoice details conform to his order). This is usually more of a spot-check deterrent against certain types of fraud than a serious verification of across-the-board accuracy. The fact that no errors are found in the sample may improve confidence a little, but does not prove that there are no errors of original record.

Errors of this type will eventually be detected when the data is used to produce an output (as when the customer complains that his bill is wrong), or they remain undetected.
b   The data may have been corrupted by a program or operating fault. The relevant tactics are discussed below under audit of the system.

Audit of the system

The auditors may conduct an **outside** audit (also called 'audit around the system'), where they treat the computer system as a black box and try to verify its accuracy from inspection of the visible evidence of input data and output results. Alternatively, they may make an **inside** audit (also called 'audit through the system'), where they inspect the computer programs and operating procedures, and dumps of files, to verify that the processes carried out on the data are correct, and that the files are in a valid state. In practice, neither of these techniques can give anywhere near conclusive evidence of accuracy.

 Auditing may take the form of pre-audit, where the auditor participates in the design procedures and gives advice on tests and controls. Alternative-ly, it may be a post-audit, where the auditor verifies that the system is operating correctly when it is in production.

 The system outputs of totals, proof lists and error lists should be designed to provide an **audit trail** which would enable an auditor, on a spot check, to:

a   pick a transaction, and check that the transaction goes through the proper processes of the system, being correctly incorporated into account balances and reports; and
b   pick an account balance or report total, and identify the transactions (and brought-forward balance, if any) that make up that balance or total.

 The auditor's objectives in detecting errors are congruent with those of the systems analyst in testing the system. Maybe there is a different emphasis in that analysts tend more to be concerned with pre-audit while auditors have continuing responsibilities for post-audit. Testing techniques which do not affect computer procedures design are dealt with in chapter 16. Testing techniques which can be built into the system design include tag-and-trace, internal controls and control totals, described below.

Tag-and-trace   With this technique, all transactions are provided with a tag field which is normally unset, but which can be set by the originator for testing. All programs are written so that whenever a transaction with a tag set is encountered, the details of the tagged transaction, and the balances or state of the master before and after using the transaction, are printed out. For testing during the construction phase, all test data can be tagged for detailed checking. After implementation, the routines remain available for spot checks.

Internal controls   Programs may check the files for internal consistency even though this would be strictly unnecessary if correctness of the programs were guaranteed. A typical example is to have a program searching a sequential master check that the records are in sequence, thereby tending to confirm that previous updates have worked correctly. Range and other validation checks on output data values is another possibility.

Control totals   Faults which corrupt financial amounts used for accounting, and other 'worst mistake' items, can be detected by control totals.

 The positioning of checks on control totals is important. To have a program report the total amounts on a file it has created is not very convincing evidence of the file's accuracy just because the output total is consistent with the inputs. A program fault may have led to corrupt output even though totals were correctly accumulated. However, to have the subsequent program, which reads in the created file, re-accumulate and report the total amounts is very convincing evidence that the first program correctly recorded the data if the total amount agrees with that expected from the inputs to the first program. Barring compensating errors, the first program cannot have added or omitted anything, nor corrupted anything that would be included in the control total. (To avoid repeated exceptions, compensating errors are ignored hereafter.) To illustrate this principle, imagine a batch system (see Figure 14.1 overleaf) where the input transactions are invoices which are to be used to update a master file of accounts, increasing the debit balance of the accounts. The control total checks on invoice amount could operate as follows.

1   The user department batches the invoices, totals the invoice amounts in each batch (giving batch totals) and logs each batch sent to the computer department. The user department totals the log.
2   The batch validation program reports (a) the total of the batches accepted by the computer and (b) details of the rejected batches. If the total of the batches accepted by the computer equals the user department log total less the amounts recorded in the log as the totals of the rejected batches, then all input batches reached the batch validation program and the invoice amounts originally recorded in the accepted batches have been correctly read by the computer. This check could be routinely carried out in the user department.
3   The transaction validation program, that reads the accepted batches, reports (a) the total of the invoices on the accepted batches file read, (b) the total of rejected invoices and (c) the total of accepted invoices. If the total (a) here agrees with the total calculated by the batch validation program at step 2 above, the batch validation program correctly recorded the invoice amounts and they were correctly read by the current program. Total (a) here should also be checked for equality to total (b) added to total (c).
4   The update program, that reads the accepted transactions and the old master, reports (a) the total of the invoices on the transaction file, (b) the total of all account balances read on the old master, (c) the total of invoices rejected for database inconsistency (e.g. no account on master) and (d) the total of all account balances written forward to the new master. Total (a) should equal the total of accepted transactions calculated by the transaction validation program at step 3 above, indicating that that program and the intermediate sort correctly recorded the invoice amounts and that they were correctly read by the present program. Total (b) should equal the total (d) calculated by the previous update of the master, indicating that the account balances were correctly recorded in the previous update and correctly read by the present program. Total (d) calculated for the new master should equal total (b) plus total (a) less total (c).

Where there is more than one type of transaction (e.g. invoices, credit notes, payments, receipts, debit adjustments, credit adjustments) it is in practice desirable to retain control totals of each type, since this helps narrow the field of search if a fault is discovered (and, as a by-product, often contributes useful information to the organisation). Similarly, it is helpful to categorise master file totals where possible. Hash totals can also be used.
Although the above explanation has been in terms of human checking, in practice it is feasible and desirable to have the computer do all the checking of control totals (except the user department's check at step 2). This will entail storing the control totals calculated by one program and having the subsequent program compare its calculated total of input with the stored total. The control totals are often conveniently stored in a special record at the end of the file they control. (This record is given a dummy record key of high values so that it preserves its position through sorts.) Programs can also reconcile their own output totals to the input totals.

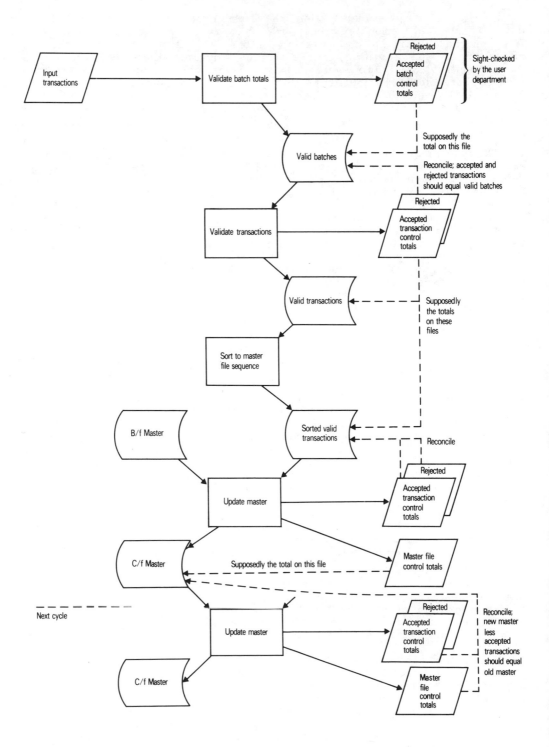

Fig. 14.1: Principles of control totals illustrated by a batch system

Failure of a control total check can then be considered an abnormal termination. However, control totals should still be printed out after each run even when they are correct, since they provide a log which helps prove the continued proper operation of the system.

Questions

1   If the control total record is placed at the end of the file it controls, rather than in a separate file of control totals, why does this give more need for procedures to ensure that the correct version of a file is mounted for a run?   (5 min)

2   With a transaction-based system, where a transaction is validated online and used to update a master file record in place, by direct access, the control total check scheme outlined above does not quite fit. Nevertheless, essentially the same controls can be established for this case. How would they work?   (10 min)

14.3 RECOVERY

The system should provide some reliable means of restoring the master and transaction files to a previous error-free state within a reasonable time after an error has been discovered or data lost or corrupted. The more reliable the means, the more recent the previous state and the more timely the restoration, so the more computer and system development resources are committed to providing for recovery. The discussion below describes a very common balance in this trade-off and supposes that catastrophes come singly. A particular case may call for different emphasis or techniques, ranging from no recovery to duplicated systems.

a   Error at source   If erroneous source data is stored on the system, there must be some way to correct it. This means that every field on a master file record should be capable of amendment by update or by deletion and reinsertion of the whole record. (Naturally, special care is needed to prevent deletion of master file records which are still current, e.g. an attempt to delete a customer account which has a non-zero balance; the update program should reject such transactions.) It also means that for every type of transaction which can update the files there should be a converse type of transaction which has the same effect but in the opposite sense. If an erroneous transaction is entered, and discovered, the converse transaction can be entered (a 'contra-entry') to negate the erroneous transaction. The proper transaction (if any) can then be submitted. As a rule, transactions should never be amended or deleted (by erasure or update-in-place) once they are entered; contra-entries and re-submissions (or, possibly, compensating transactions) should be used. In this way, a possible anomaly of having a transaction altered after it has been used for reporting is avoided, and transaction files accurately show the transaction entries originally raised at source, preserving the audit trail.
  An exception may arise in the case of a 'suspense' or 'holding' transaction file which is deemed by the users to hold details of transactions which are not yet 'concluded'. Update or deletion of such transactions, prior to their being concluded, is legitimate. However, once the transactions are concluded, an event usually signalled by their use to update a master file or to supply a figure which is incorporated into the accounts of the organisation, contra-entries or compensatory transactions should be used for corrections.
b   Other errors   For this discussion, it will be helpful to define the following terms.

  Current file: the file (possibly in the course of being updated) which
    is intended to reflect the present desired status of the data;

Backup file: an earlier edition or copy of the file which can be used
to restore an earlier status of the data;

Differential file: a file which can be used to update a backup file to
current, or later backup, status;

Residual differential: the unsecured transactions which, after the best
recovery possible has been made following loss or corruption of data,
need to be re-entered at source to restore the current file.

The discussion is developed by considering (1) the case of the file which
is updated by copying forward, and (2) the case of the file updated in
place.

1   The copy-forward file   When a master is updated by copying forward,
the earlier edition forms a natural backup and the transaction file(s) form
a natural differential file. Should a fault occur while the current file is
being created, with the result that the current file (alone) is corrupt, the
update program can simply be started again from the beginning using the
same inputs.

Should the fault corrupt the input master (the 'father') as well as the
current file being output (the 'son'), the current file will be irrecoverable
unless a backup input master can be made available. If the still earlier
edition of the master (the 'grandfather') and the differential files between
grandfather and father have been saved, the father can be recovered by
re-running the update with those files. (It is often decided to store the
media containing grandfather files and their differential files at a separate
location to reduce the effect of a catastrophic loss such as a fire at the
computer centre. As the grandfather files become great-grandfathers, so the
media are released for re-use.) The re-running of the update must be done
with the version of the program and the input parameters that prevailed
at the time of the original update.

Should the fault corrupt the input transaction file(s) as well as the current
file being output, the current file will be irrecoverable unless a backup
transaction file can be made available. At least one backup copy of the
current transactions should be retained until the update is successfully
completed. Such a backup copy may exist as a by-product of earlier
processing, e.g. a file of unsorted transactions or unvalidated transactions.
By re-processing the backup, the current transaction file can be restored.

Naturally, should recovery by using standard application programs produce
large quantities of printed or other output which is not required, it will
be desirable if there is the option of switching this off.

Turning now to the less likely possibility of corruption of the files while
they are not being used, the worst case will occur if a catastrophe destroys
the father, the father-son differentials, the son (current master) and the
current transactions while they are all at the same site. The remote
grandfather and grandfather-father differential can be processed to restore
the father, but there is a residual differential between the father and son
as well as a lost current transaction file. Consideration should be given
to (a) the need to remove the transaction files to a remote location as soon
as possible after processing, (b) the need in user departments to retain
source information about transactions entered during the current and
preceding processing cycles, so that the residual differential can be resubmitted if necessary, and (c) the need for shorter processing cycles to
reduce the size of the residual differential.

It should not be overlooked that in this grandfather-father-son technique
the backup files occur as an incidental by-product of copying forward, with
the frequency dictated by other considerations such as accounting periods
or pay frequency. Should it be judged that the backup cycle is too frequent,
it can be slowed down by increasing the differential held in the remote
location and leaving out intermediate generation backups. For example, with
a daily update, the start-of-week masters could be stored in the remote

location, followed by each day's transactions. The only likely reason for judging that a backup cycle is too infrequent is that the master file or transaction files are so large that going back to the beginning in the event of mid-program failure may be too large a task to contemplate normally. (Of course, loss or corruption of the files while they are not in use would also involve a large effort in recovery, but this is usually thought to be too long a shot to warrant special action in a batch system. Procedures to reduce recovery effort are further discussed in connection with online systems, below.) Recovery from mid-program failure can be alleviated by using operating system facilities to take checkpoint dumps of program and file status at strategic points (e.g. end of tape reel), permitting processing to be re-started from that point should it prove necessary (see the RERUN options in standard COBOL).

2   **Update-in-place**   With a batch update-in-place, a backup copy of the master file may be dumped at suitable intervals, e.g. start of day, start of week. The batches of transactions again provide the differential files. Grandfather-father-son security can be achieved by removing a previous period's backup file and differential files to a remote location. Should a recovery be necessary, it may be desirable to merge and sort the trans-actions before using them to re-create the master. Consideration should then be given to the problem of preserving the proper original sequence of merged differential files, for example by sorting to date or transaction number within the key of the master. For an even faster recovery, copies of those master records updated can be dumped onto a differential file during update (the copy showing the updated state of the master). The most recent versions of each master record can then be selected from the differential file and merged with the backup master.

When a backup copy of an indexed master is made, this is often an opportune time to make the copy by reading the records in key sequence and then to reload the original master from the backup copy, reorganising the file for greater efficiency by eliminating overflow records. If the copy is made to a direct access medium, it may be possible to swap the backup with the original, instead.

With an online update-in-place, it is usually especially necessary to restore the system quickly. At the same time, there are no natural by-product backup copies of transactions entered. The basic recovery philosophy is usually to copy the master at fixed intervals and then to create an artificial 'batch' – a differential file – by dumping subsequent transactions as they are entered. The differential file created, the transaction log, is in any case desired to preserve the audit trail and to provide users with a permanent record of accepted transactions. At the end of each processing cycle, the log can be printed out, together with control totals, the report being divided into transaction types within user. Clerical procedures should check out the reported totals with those entered at source. The master file copies should be taken, and the log file closed and re-opened, at clearly identifiable points of time, such as end-of-day, to assist users in re-submitting the residual differential should recovery be necessary when the current log is lost.

With a large master and a busy online system, recovery from the trans-action log may be too slow to contemplate for routine system failures, or frequent dumps of the master may be too heavy on system resources. The use of a log of updated master records, in addition to the transaction log, may go some way to alleviating this problem, but the effect on system performance of the additional data transfers this involves should be investigated. Other techniques are surveyed in Verhofstad (1).

Unscheduled interruptions to processing leave special problems with online updating-in-place. With an offline batch system recall, recovery is made to the point at the opening of the program or at the last checkpoint. The batch of transactions, or the balance of the batch beyond the checkpoint, is reprocessed. With an online system, the interruption may leave the master and differential files improperly closed, and there may be untransferred records in system buffers. (To be strictly accurate, the same things may happen with a batch system. The difference is that with a batch system the source transactions are preserved in machine-readable form and recovery is rather easier.) The operating system may help with these problems (by scavenging the buffers and saving the files). Failing this, the analyst will need to arrange for frequent dump-differential cycles or frequent checkpoints, to minimise the residual differential. (If the updated masters are copied on a log, the originals being left untouched, and checkpoints are taken periodically, re-start procedures can very quickly restore the master and transaction logs to their checkpoint state. A penalty is that application programs will have to check first in the master log to make sure they get the latest version of masters. There is much to be said for letting the user declare a personal checkpoint periodically at the terminal and giving him the right to restore the system to checkpoint status following a failure. The residual differential is then those transactions entered since the last check-point.)

Questions

1   How do integrity considerations affect device assignments?   (2 min)

2   If control totals are stored on records not included in the file they control, what additional recovery procedures are necessary?   (2 min)

3   Referring to the answer to section 13.3, question 3, how has tuning in this fashion complicated recovery?   (5 min)

4   When a file has been split (section 11.3), if only one part is lost do you need to recover both parts?   (5 min)

ASSIGNMENT

Continuing the assignment at the end of chapter 10, complete the database definition, computer run design, integrity and security procedures and program specification for the mail-order wine club order entry system.

ANSWER POINTERS

Section 14.1

1   Internal: accounting system – consider insider involvement as partners;
           non-accounting system – consider outsider consultation.
     External: outsider consultation; send all documentation for scrutiny.
  The auditors should be involved with non-accounting systems because such systems could corrupt or provide illicit access to accounting systems.

2   Not if the computer operator is able to ignore system warnings about incorrect files or to by-pass label checking. The possibility of accidental use of spurious versions of files produced during program testing should also be explored.

Section 14.2

1   A current control total recorded separately from the file is also a check that the right file was mounted (barring an unlikely coincidence).

2   Users log and total transactions manually. Accepted transactions are logged by the computer as well as being used to update the master. When the master is opened, the previously-stored control totals are accessed. The

control totals are updated in memory. When the master is closed, the contol totals are stored.
 The total of the logged transactions is reported to the users for checking.
 The master is read, throughout, and the accumulated totals compared with the control totals. (This can be combined with copying for recovery by reading the records in key sequence – see section 14.3.)

Section 14.3

1   Choose the most reliable device for sensitive files. Hedge your bets by allocating current master, backup master and differential files to different devices, to reduce the consequence of device malfunction.

2   Need to recover control totals as well as file.

3   No by-product differential file. If original inputs are large-volume, slow-to-read (e.g. OCR), recovery may be slow and storage of differential bulky.

4   Not if there was no residual differential (although the update procedures may require you to recover and update other parts if they are designed to enforce joint update as a discipline and there is no way to over-ride this). If there was a residual differential, the re-input transactions may not be quite equal to the original ones, so in this case it is obligatory to recover both parts if the risk of anomaly is to be avoided.

REFERENCES

(1) Verhofstad, J. S. M., Recovery techniques for database systems, **Computing Surveys**, 10, 2, 167–95, June 1978. 43 refs.
(2) Diroff, T. F., The protection of computer facilities and equipment: physical security, **Data Base**, Volume 10, pp. 15–24, Summer 1978.
(3) Gibbons, T. K., **Integrity and recovery in computer systems**, NCC Publications, 1976.
(4) Farr, A. A., Chadwick, B., and Wong, K. K., **Security for computer systems**, NCC Publications, 1972.
(5) The security of managers' information, **EDP Analyzer**, 17, 7, July 1979.

15   Reporting and training

15.1 WRITTEN REPORTS

Comprehensive suggestions for desired reports, and the contents of them, are given in the book, already cited, by Hice, Turner and Cashwell and in the NCC Data Processing Documentation Standards. In this chapter, the aim is only to highlight a few aspects which are shown by experience to be especially important or which give rise to difficulty.

Getting started   If you feel a long time is being spent without getting started on the report, or without making progress on it, try:

going where you will be free from interruption;
driving out distracting thoughts, or giving in to the distraction whole-heartedly, returning to your writing at a better time;
reviewing the relevant system documentation or other notes you have made so far;
boning up on the subject;
writing a table of contents;
writing first about the things that interest you most;
being systematic.

Being systematic   Ask yourself:

who are the recipients of my report?
considering each type of recipient, am I trying to get a reaction from them, approval for example, or am I just informing them, or am I trying to instruct them, or supply them with a reference manual?
if there is a mixed readership, should I write separate reports for each type of reader, or structure the report with different levels of detail or specialist sections and appendices?
what will be the surest way of getting the desired reaction? what will interest the readers most?
what do I need to demonstrate, what facts must I marshal, what do I need to find out and present?
what is the logical order to develop the argument?

Style   Use your natural style and everyday words. Read the delightful book by Gowers (1); the ideal time to do this is just after you have finished your first draft of the report. If you use short sentences and short, simple words, you run less risk of submitting your reader to the mental torture of unravelling a complicated sentence and less risk of accidentally obscuring your meaning – but nearly every rule about style was made to be broken, and it is certainly possible to get away with sentences of enormous length providing a little care is taken to present the clauses of the sentence in step-by-step fashion, supporting them with suitable punctuation.
  Some reports seem to be written as though the choice of words was unimportant. It's as if such a report writer believes that providing **he** knows what he means, the reader will somehow pick up the message, no matter how badly expressed. There is a grain of truth here, but nevertheless this philosophy is a mistaken one. The grain of truth is that in addition to the impersonal sense of the words (the objective facts and ideas they

cxpress), meaning is passed by:

the context, which is understood by the reader;

the emotive overtones of the words, which tend to show how the author feels about the facts and ideas;

the written equivalent of 'tone of voice', which tends to reveal the author's attitude to the reader.

However, if the words do not have objective meaning, the writer can have passed on no new ideas or facts, since the sense of the message rests on the reader's prior knowledge. At most, the author can have communicated only emotions and attitudes, and drawn the reader's attention to something he already knew. If a message is worth communicating at all, it is worth the effort of searching for a clear and explicit form of words. Just as important, the search for the right words can sharpen up the writer's own ideas about his subject.

Every programmer knows that he can hardly expect to write the source code of a large program and get it right first time, especially if he does not spend sufficient effort on planning the program beforehand. Writing a report in clear English is quite as error-prone as writing a computer program, and there is no compiler to detect the errors of syntax. The experienced programmer's discipline of thoughtful planning, attention to detail, desk-checking and inspection by one's peers can be applied with equal validity to any large prose text.

While researching for this section, I came across directly contradictory advice: (a) use an impersonal style; (b) prefer the personal to the impersonal. There are many people who prefer to write impersonally, presumably because this preserves a tone of objectivity which is fitting for an impartial report. Against this, the author who avoids first and second persons runs a greater risk of obscuring his meaning. This is because sentences phrased impersonally tend to be more circuitous or unfamiliar, and often give rise to the more complicated passive voice instead of the simple active voice. Preserving clarity under these circumstances calls for even greater skill and effort. On the other hand, a passage which contains a forest of 'I's and 'me's may give an undesired impression of egocentrism, a consequence which has driven many a writer to invent for himself a fictitious plural personality so that he can write with neutral 'we' and 'us'. I like the slight extra warmth that comes with the personal style, and the way it tends to break down an unjustified aura of objectivity, so reminding the reader of the human frailty that lies behind the coldly-printed words.

Terminology Buzzwords, coined words and unfamiliar language pervade the literature of computing, even when the usage is avoidable and the meaning shared by hardly any of the readership. Examples can be found any week in the popular computing press, where too much use is made of unexplained acronyms and cryptic references, with the apparent intention of impressing on the reader his own ignorance – or perhaps to flatter him by suggesting that he is a member of the cognoscenti. Such usages do not impress me, for one, and only cause the offending article to be quickly discarded. Other bad examples can be found in technical journals, but here the game often seems to be not to make one's point too plain and understandable, lest ease of understanding should be thought to diminish the quality of the ideas.

Questions

1 Does it matter if grammar is wrong?   (3 min)

2 The Fog Index of a passage of text is calculated by adding
   100 times the fraction of words which exceed two syllables
        to
   40% of the average number of words in a sentence.

An index value of 15 is reckoned to be normal. A value of 10 is typical of the clipped style of popular-press journalism. Heavy, difficult reading may give rise to a value of 20 or more. What is the Fog Index of this question? Can the Fog Index be taken seriously?   (10 min)

15.2 PRESENTATIONS

Assuming you wish the recipients to read and take action on your report, it should be handed to them personally (or at least heralded by a telephone call) and the nature of the desired action explained. The deadline, and the reason for it, should be stated at the same time. The recipient who feels personally involved in reading the report, and who understands the urgency of the deadline, is likely to rank it high among his priorities. The reader who has impersonally received the report, and who believes the deadline is artificial, will find it easier to neglect.

Even an excellently-produced written report cannot match the impact of a live presentation, and it can produce misunderstanding or misplaced emphases. Even the most conscientious manager may be unable to give a written report the scrutiny hoped for by the author, because of competing claims for the manager's attention. Even the keenest readers may be unwilling to act until they have discussed the features of the report amongst themselves. These facts argue for making a more substantial presentation to the intended recipients, as a group. If involvement has been encouraged, there will be few surprises for the audience at this presentation. It will be more of a chance for them to review the whole plan and reach a consensus. Those who have contributed should be encouraged to participate in the presentation as well.

A first presentation is a nerve-racking experience for most people. It takes a long time to develop a refined technique. The three worst risks for the novice, assuming he knows his subject, are those of drying up, speaking too fast, and speaking for too long.

Here is a four-point plan to help with a first presentation.

1   Prepare a summary for the start and the end   A prepared summary will ensure that there is a clear message to be given, reducing the chance of freezing up at the outset. Summarising helps sharpen ideas about what are the most important things to say. The opening summary arouses interest and gives a plan of the presentation, onto which the audience can map what follows, aiding their comprehension. The closing summary should reinforce the main ideas and call for the desired action.

People remember rather little of what is said in a verbal presentation. A certain amount of repetition is essential if you want to get your points remembered. Schoolteachers have a saying which goes 'Tell'em what you're going to tell'em, then tell'em, then tell'em what you told'em'. The summaries should contain the key facts you want your audience to recall.

2   Keep it short   Twenty minutes of lecture-style presentation is about the most an audience can pay attention to. If you need longer to make your case, you must find some way of providing a break – or perhaps you need two presentations. When people over-run, they are often making too much play of secondary points which dilute their main arguments and reduce their impact. In any longish presentation, the threads of the talk should be gathered together at intervals by reminding the audience where you have come from in your argument and where you are now going.

3   Take it slowly   The preparation for, and anticipation of, the present-ation can build up a certain tension in the presenter. A small amount of anxiety is desirable, since it helps keep the presenter on his toes. But often, once the hurdle of getting started is over, the flood-gates holding back the tide of ideas are opened, and whoosh! – words surge over the audience, who quickly drown, and the reservoir is emptied. Sometimes, perhaps as a result of preparing too much material, the presenter realises he is not going to get through it – and his response is to go faster, when it should be to discard inessentials. Sometimes it seems that the presenter believes all silences are awkward ones and it is his duty to maintain a continuous noise which will fill every fleeting moment.

Obviously, it is also possible to go too slowly, so what is desired is a measured pace. Use brief pauses for effect after each main point. If you know acceleration is your problem, write 'TAKE IT SLOWLY' on the top of each page of your notes. At least this will give the audience the odd chance of drawing breath.

4   Use visuals   It' is a rare person who has the charisma and oratory needed to carry off a presentation without visual aids. Handouts, specimens, flip-charts, overhead projector foils, and so on, make talks more enjoyable and, to the extent that they 'repeat' the words, provide another means of reinforcement.

Overhead projector foils provide the means to make effective presentations with an economy of effort. Avoid crowding too much on a foil; one bold diagram or about five items or points is the usual maximum. The essential arguments may be watered down if too many foils are prepared. It is easily feasible to give a twenty-minute talk centred around one foil. Four or five foils is the most that should normally be considered for such a period.

No matter how engrossing the talk, people listening to a presentation have short periods when they lose attention and hear nothing of what is said – 'microsleeps', usually lasting only a few seconds. Visual aids which repeat the spoken words, and reveal the speaker's plan, help the 'sleeper' to pick up his lost thread.

15.3   GUIDES AND MANUALS

The idea behind a user guide or manual is that it is a document the user can turn to, for learning a new procedure or overcoming a difficulty in operating a procedure. In practice, though, few users choose to learn or overcome difficulties in this way. They prefer to find a human being who can help them. Consideration should be given to providing human user 'guides' and points of reference.

With interactive systems, it is valuable to supply the users with a 'HELP' command. Few systems will justify the expense of developing computer guidance to overcome difficulties, but perhaps a HELP request could be answered by a human firefighter.

Although written guides and manuals may be insufficient, they are usually still necessary even if only to provide a reference of last resort. The following steps are suggested.

1   Identify the purpose   To train by self-instruction? To back up a training course? To be an everyday job aid, for example providing codes and menus? To help overcome routine on-the-job problems? To prescribe exceptional procedures, e.g. for recovery after disaster?

2   Prepare a draft   Bear in mind the purpose. A work of reference will need to be comprehensive, precise and thoughtfully indexed. A training text will need more examples and exercises.

3   Test the draft   Try it out on people who are representative of the target readership, if possible. Monitor their difficulties.

4   Revise   Work out the reason for difficulties, listen to comments, and revise the guide in the light of this feedback.

A particular problem in guides and manuals is the explanation of complex rules or procedures. It is easy to get tongue-tied when describing an involved procedure or set of rules. A flowchart, Nassi-Shneiderman chart (see Figure 12.6) or decision table can help. If a narrative is still desired, writing the narrative after having prepared a chart or decision table will often lead to a clearer explanation.

Exercise

1   There follows an extract from a Department of Health and Social Security leaflet issued February 1978, giving national insurance guidance for married women. The leaflet is intended to give general guidance only and not to be treated as a complete and authoritative statement of the law. A married woman who, on 5th April 1978, had an established right to pay reduced rate contributions may, if she wished, retain that right. Alternatively, she could choose 'full liability', i.e. elect to pay standard rate contributions.
  Prepare a decision table which would help an employed married woman to find out the amount of her contributions. Prepare another table which would help any person to discover which other leaflets contain further information for them. Make a note of any ambiguity you discover while doing this and prepare a precise statement of the nature of the ambiguity. Finally, making your own assumptions about the most reasonable interpretation of any ambiguity, redraft the extract.

CLASSES OF CONTRIBUTIONS

Class 1 contributions are paid by you and your employer when you are employed and your earnings reach a specified minimum limit (for the 1978/79 tax year this is £17.50 a week). These contributions are related to earnings – so the more you earn up to a specified upper limit (£120 a week for the 1978/79 tax year) the more you pay.

The contribution is a percentage of all your earnings up to the upper limit and the rate is fixed towards the end of each calendar year, to operate from the start of the following tax year.

The standard rate for 1978/79 is 6.5% and the reduced rate is 2% of all earnings up to the upper earnings limit. However if you are in contracted-out employment you pay $2\frac{1}{2}$% less than the standard-rate on your earnings between these limits. The contribution rate for women with reduced liability is 2% of all earnings up to the upper limit.

If you work for more than one employer, Class 1 contributions have to be paid in each job in which your earnings reach the minimum limit, unless you have arranged to defer payment of some of your contributions. If you pay more contributions than a set amount for the year the excess will be refunded.

Further information: Leaflets N140 (Contributions for employees) and NP28 (Contributions for people with more than one job).

Class 2 contributions (£1.90 a week in 1978/79) are normally payable if you are self-employed, but you may have chosen not to pay these contributions at all – see paragraph 5. Alternatively, if you expect your earnings from self-employment to be low, you can apply to be excused payment of Class 2 contributions.

Further information: Leaflets NI41 (Guidance for the self-employed) and NI27A (Guidance for people with small earnings from self-employment).

Class 3 contributions (£1.80 a week in 1978/79) are voluntary and can help you to qualify for certain benefits, but only if the Class 1 or Class 2 contributions you have paid or been credited with are not enough. You cannot pay Class 3 contributions for any tax year for which you have chosen reduced liability – see paragraph 5.

Further information: Leaflet NI42 (Class 3).

Class 4 contributions are payable on demand to the Inland Revenue and collected with Schedule D income tax if you earn more than a certain figure a year from self-employment (for 1978/79, 5% of profits or gains between £2000 and £6250 a year).

Further information: Leaflet NP18 (Class 4).

If you work for an employer as well as being self-employed, you may be liable for Class 1 and Class 2 contributions and, if appropriate, Class 4 contributions as well, if you have not arranged to defer payment of some of your contributions.

Further information: Leaflet NP18 (Class 4).

Current contribution rates and the earnings limits are always given in the latest edition of leaflet NI208.

15.4 TRAINING

The first temptation the trainer must resist is the thought that he is there to teach the learner to 'understand' or to 'appreciate' something new, or similar vague targets which are not expressed in terms of clearly observable behaviour. The question the trainer must ask himself is 'How will I know that he understands, appreciates, etc.?' The trainee's head cannot be opened up to look inside, so how will successful training be recognised? Reframing the objectives in terms of **target behaviours** of the trainee suggests ways of achieving the objectives, and also ways of testing the attainment of the trainee and the effectiveness of the training methods.

For example, the simplest type of objective is one of imparting **knowledge**. Here, the observable behaviour is **recall** of the facts. The instructor should define the knowledge behaviours in terms of specific facts which are to be recalled. Thus, not 'the trainee should know the coding system' but 'given any department name, the trainee should be able to recall the department code, and vice versa'.

A different type of objective is **comprehension**. Here the observable behaviour is that the trainee can both recall the facts and describe or illustrate them using words, actions or examples which are different from those the instructor used. So a target should not be 'to understand the highway code', but instead a series of objectives such as 'to be able to give his own examples of situations where a driver should not overtake'.

A common objective is that of **application**. Here the trainee should be able to apply the ideas, which he has comprehended, to fresh cases. The fresh cases are not identical to any of those met in instruction, but are of the same general type. An application target should not say 'to appreciate the company's rules for giving credit' but rather 'given the customer's time and cash credit limit, and aged account balances, to decide whether or not to satisfy a request for a stated further amount of credit'.

Armed with a comprehensive list of well-defined target behaviours, the trainer can decide whether or not to adopt on-the-job or classroom-style methods, and whether the most effective technique will be lecture, tutorial, discussion group, casework, example, simulation, exercise, rehearsal, practice, reading, programmed text, etc. or a mixture of them. Many of the ideas of sections 15.1–15.3 on report writing, making presentations and preparing guides can be applied in constructing a course.

Each training course should end with an evaluation of the instructor and the course by the trainees, so that instruction methods can be reviewed. There should also be an evaluation of the proficiency of the trainees, by monitoring or examining how well they meet the behavioural objectives. The effectiveness of the training can then be established and, if the particular course is to be re-used, instruction methods can be improved next time round.

ASSIGNMENT

Continuing the assignment at the end of chapter 14, prepare user guides for the mail-order wine club order processing system you have designed. Prepare a plan of the training and target behaviours for the order clerks and pickers.

ANSWER POINTERS

Section 15.1

1   It is perfectly feasible for an ungrammatical report to be understood. If the writer means his report to be effective, a motivator, and the reader (rightly or wrongly) considers that grammar matters, then grammar matters.

2   I made the Fog Index 16, but there were no instructions given how to treat the numbers, the percentage and the hyphenated word. The layout of the first sentence made it very much clearer than it would have been as a continuous string of words, but the Fog Index does not take account of this. Although every COBOL programmer is quite clear what makes a 'sentence', things are not so cut-and-dried in natural language. One person's semi-colon is another person's full stop. There are thus face-value objections to the reliablity of the Fog Index.

A further objection is that it is not stated whether a high or low Fog Index is a good thing or a bad thing. There is not much point in working out the index value if there is no action prescribed. The implication is that reading with a high Fog Index is generally less easy to understand; but who wants a meal of infant children's books all the time?

There is clearly an element of truth in the proposition that long words and long sentences tend to increase reading difficulty. The trouble is, there seem to be other components of difficulty with similar importance. (In the text given as an exercise at the end of section 15.3, for example, the difficulty in extracting the meaning has rather little to do with sentence length. In fact, most of the sentences seem to me to be constructed quite expertly. Placing oneself in the shoes of a person reading the text for guidance, the difficulty stems at least in part from not being sure which parts of the text can be ignored, as inapplicable, and which parts have to be retained mentally, to support the interpretation of other parts which are applicable. A person reading it is quite likely not to know whether they are in contracted-out employment, or not; whether they have reduced liability, or not; whether the earnings they expect from self-employment are 'low' or not; whether their past Class 1 or Class 2 contributions were enough to qualify for the 'certain benefits', or not. The person who will understand the guidance most easily is a sort of model citizen who knows his personal status in the terms defined by the administration; someone, one suspects, who is rather like the administrator.) I would not let the Fog Index guide my style; but if I find I have written a difficult sentence, I take the lesson from it and try substituting shorter words or splitting the sentence in two.

REFERENCES

(1) Gowers, Sir E., **The complete plain words**, Pelican, 1973.
(2) Vargas, J. S., **Writing worthwhile behavioural objectives**, Harper–Row, 1972.

16 Testing and handover

16.1 PLANNING FOR CONVERSION

Whenever there is a file of data, such as a customer master file or commodity master file, which must be available to support the new system operation, there will be a need to develop a conversion system, as well as the operational system, if the files do not already exist in the right format. This file conversion cannot usually be planned successfully as a last-minute afterthought, but should be given the same attention as the operational system (see also chapter 2). Figure 16.1 illustrates a typical case: a 'one-shot' approach to converting master files. After they have been converted in anticipation of full-scale live acceptance testing, the planned master file maintenance procedures operate to keep them up-to-date. Sometimes it can help to do a one-shot conversion earlier, before the system test. In this case, maintenance procedures must be ready and tested before conversion and used to keep the masters up-to-date during the system test (as well as thereafter).

With a large-volume master which exists only in manual form, or which requires manual corrections of error or format, the work involved in a one-shot conversion may be too great. A solution to this is to do an 'as required' conversion. Here, master records are inserted only as and when required to support the transactions that occur when the new system is operational. For example, consider a system where a large customer account master is needed to permit updating by invoice transactions. As each invoice is raised, a manual check can be made to see whether or not customer details have already been submitted and, if they have not, to submit these along with the invoice. The effect is to spread the clerical workload over a period of time. At the outset every transaction will need an accompanying master insertion, but gradually the master file will become complete.

Questions

1   The main disadvantage of an 'as required' conversion is that some activities may require both manual and computer work throughout the conversion period. There is also an annoying overhead in checking whether or not the master record already exists. How can these disadvantages be alleviated?   (3 min)

2   Apart from computer hardware and programs, what needs to be ready to train users in preparation for final testing and handover of an inter-active system?   (15 min)

3   Who should do the user training?   (3 min)

4   What documents will be needed to support computer operations and system maintenance?   (10 min)

16.2 SYSTEM TESTING

In this section, testing is considered short of operating the system live in

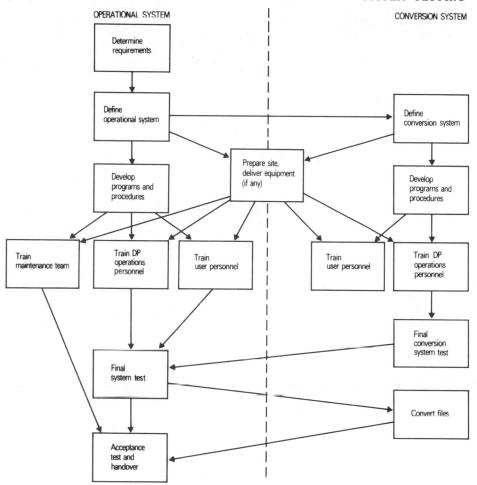

OPERATIONAL SYSTEM

CONVERSION SYSTEM

Fig. 16.1: Typical preparations for conversion and implementation

the real environment (see section 16.3).
Testing is considered from the points of view of:

functional testing; performance testing; fallback and recovery testing,
quality assurance; regression testing.

Functional testing   This seeks to find functional faults in programs and
procedures which give rise to erroneous output from the system. The
recommended tactics are (a) thoughtful selection of test cases for submission,
and (b) inspection. With a system where freedom from error is paramount,
consideration could also be given to duplication of the system development
by independent design and construction ('dual programming'), with a view
to comparing the results of the two systems.
Test cases are submitted to seek out hypothetical bugs. Programmers,
systems analysts and users may all hypothesise different types of bug, or
take different views of the vulnerability of the system to different types
of error. All three should submit test cases.
Programmers should test the modules/programs they have written such that:

every statement which can be executed (as a result of input values

which are not excluded from the data) is executed at least once;
every possible condition in the program is executed in each of its senses;
the paths between conditions are explored, as far as can be judged, with the minimum and maximum possible values of the variables referenced in the preceding condition.

These tests tend to confirm both the accuracy and the robustness of the program's functions. They prove little, of course, if the specification is wrong or the programmer has misinterpreted it.

Analysts should also test the programs, or suites of programs, by submitting test cases which in their experience are likely to expose faults. Such cases could include important combinations of valid and invalid input values, significant values and quantities of data, combinations of end-of-file conditions, zero values, zero quantities and so on.

Users should test the computer systems and clerical/operational procedures by submitting test cases which are significant to them. They should be encouraged to submit exceptional test cases as well as normal cases. All the cycles of the system should be tested.

The result of each test case above should have its accuracy certified by the tester. There is no point in submitting thoughtfully-selected test cases if each one is not checked. Checking of results which do not normally appear on designed outputs is aided by tag-and-trace routines (see section 14.2).

It is also reasonable to volume-test the functions by submitting large quantities of test cases in the hope of 'blowing up' the programs (causing abnormal termination). These cases may be taken from live files or produced by a test data generator. Successful handling of the volume test, though, adds rather little to confidence compared with the thoughtfully-selected cases.

The other main testing philosophy is to scrutinise the documented procedures and programs to try to judge their correctness. Sometimes a 'walkthrough' committee is established to review all the system documentation, including the test case results. As far as functional testing of individual programs is concerned, detailed inspection of the source code by a knowledgeable person (not the original programmer) is probably more effective than review by a committee.

Performance testing   Here the objective is to discover how well the system meets the organisation's targets and other measures of effectiveness and design aims such as execution time, response time, average time to process a given type of transaction, etc. This performance testing is greatly aided if the measures of effectiveness are identified and agreed at the outset and if evaluation aids are built into the system where necessary, e.g. reports of execution times, a calculated average response time, a count of the number of transactions by type, etc. See also section 2.3, question 2, and Systems Management for an enlargement of this topic.

Of course, the effectiveness of the system will also be of concern during acceptance testing and after implementation. The objective at the present stage is, through simulated working and test cases, to give early warning of performance weaknesses. Corrective action can then be instituted if necessary, and simple fine tuning of procedures, displays, report layouts can be done. The volume-test of functions can often also serve for the performance test.

Fallback and recovery testing   There is not much point to making fallback and recovery plans if they do not work when they are needed – as is quite likely to be the case if they have not been tested. The designed recovery procedures should be checked out, including running on the standby computer if this is part of recovery plans. Recovery from a simulated catastrophe (e.g. complete destruction of the building housing the computer centre) should be explored. The time to discover a weakness in the plan is **not** when the disaster has occurred; although, as with most things, the effort and expense of prevention must be weighed up with the risk of the threat occurring and the extent of the damage if it does.

Because there are often hidden snags with using standby computers at remote sites, it is best to plan for an early test to allow revision before the new system relies upon the standby arrangements.

Quality assurance   This seeks to verify that the delivered system meets the organisation expectations for quality and standards. The adequacy and content of the documentation, the testing done and implementation plans should be reviewed by a walkthrough committee of high calibre personnel.

Regression testing   When a system is amended, there is sometimes an unexpected side-effect which corrupts a non-amended part of the system. Regression testing seeks to confirm the continued correct functions or performances of the system following a change.

Question

1   How would you make regression testing easier?   (5 min)

16.3   ACCEPTANCE TEST AND HANDOVER

Acceptance testing starts when it is believed that the system is ready for live operation. It is the users' last chance to verify that the system meets their requirements, before they finally rely upon it.

Sometimes, users conduct a pre-operation 'one-shot' acceptance test before an immediate handover of the new system and cutover of working from the old system to the new. This is a quick and effective method – if all goes well. It is also very risky. It should not be considered unless there is an effective standby system or reliability is not important. The recommended, and more cautious, approach is for a phased handover or phased cutover, or a combination of the two.

Phased handover requires that the old system and the new system are run in parallel for some period. During this period, the new system is monitored against the old and the discrepancies in the results are reconciled, or faults in the new system are fixed, or – as sometimes happens   a point is reached where more confidence is held in the results of the new system than those of the old, even though unexplained discrepancies remain.

As the different functions of the new system pass their acceptance tests, so there can be cutover to that function, dropping the old. For example, the data entry procedures may initially be used on both new and old systems. When the data capture portion of the system is accepted, the old system data capture can stop. Other old system procedures, e.g. report production and bookkeeping, continue until they are accepted in turn, using for example the proof lists and control totals produced by the new system. In this way, the old system functions are phased out one by one.

Phased cutover entails accepting the system for only a portion of the transactions it is designed for, limiting the effect of possible faults. For example, a particular unit may undertake a pilot operation, e.g. one branch in a multi-branch organisation transfers to the new system. A particular class of business or type of transaction may be chosen for cutover, leaving the others until the first is checked out and accepted. Phased cutover is best used in conjunction with phased handover.

Questions

1   There is often a 'political' pressure to accelerate handover in an organisation, and analysts (usually mistakenly) sometimes go along with such a hastened implementation. Why do you think this is?   (10 min)

2   Define the difference between handover and cutover as used above.

(5 min)

16.4 MAINTENANCE

Maintenance is considered from the following viewpoints:

> request for change;
> firefighting;
> documentation control.

Request for change   By the time of handover, amendments to the system will probably already be required (see section 2.2, question 2). Further improvements to the system, or adaptations to meet changed circumstances, are bound to be needed after it is operational. During the first two years of a system's life, maintenance costs typically amount to a third of the original development costs; possibly more if one takes into account that difficulty in maintaining a system may lead to it being discarded and completely redesigned earlier than might otherwise be the case.

Each amendment should be handled like a 'mini-project'. The expense of the change should be justified, the change should be planned and controlled, testing, retraining etc. should be carried out as necessary to preserve the standards of the system. To keep the procedure simple, a change is often approved on a short form describing the nature of the change, its purpose and its costs and benefits if these are at issue (see, for example, the NCC Amendment Notification). The user, sponsor and analyst should endorse the change request.

Change requests should be assigned priorities, and personnel should be assigned to implementing the changes in the light of these priorities.

Another short form may be used to permit simple planning and monitoring of the change. This will contain a list of the tasks associated with the change, together with a target date and responsible person for each task. Completion dates and references to documentation amendments can be recorded on the form (see the NCC Amendment Log).

Firefighting   In spite of all efforts, emergencies will arise when an unsuspected bug shows up by causing an error in an operational program. It is very tempting to shout 'fire' and have all hands running to the pumps. This gets everyone working frenziedly to rectify the fault at all costs. The problem with this enthusiastic response is that documentation of the change may be sketchy or even absent. The participants themselves sometimes cannot even recall exactly how the change was effected, especially if it was a deep-seated bug. This leaves the system and documentation in an unsatisfactory state for continued operation.

Keeping a cool head, one should ask 'Is there a way I can solve this emergency and fix the bug through the regular change procedure?' Maybe the fault is exposed by one particular transaction. Perhaps the system can be recovered to a state prior to that transaction, the transaction can be removed and then the system can be recovered to current state minus the transaction. If the processing of the transaction cannot wait until the amendment to the system is made, perhaps it can be processed manually on the fallback system, or perhaps a procedure can be devised forthwith to cater for the transaction manually. These choices are less risky than emergency changes.

Often, a major problem is diagnosis of the fault, let alone correction of it. Diagnosis is an activity that is more suitable for a crash effort. The problems may come if the diagnosticians go on to create instant solutions. A crash effort should, therefore, culminate first in a diagnostic report alone. Then a calm decision can be made about a solution.

If it should be decided, none the less, to implement an emergency change, the regular change procedure should be carried out in retrospect. This will help all concerned to learn from the experience as well as improving the quality of the documentation.

Documentation control   As a result of the amendment, some of the documentation of the system will be placed in the system history file (see section 6.1) and replaced by new documents. A problem may arise if copies of the original document have been circulated, e.g. in a user manual. It may be necessary to update all such copies.

The problem for the maintenance team is to know whether or not the original document was copied and, if so, where the copies were circulated (e.g. in which reports). The first point may be answered by the analysts adopting a convention, whenever a document is copied, of marking the original distinctively. The second point can be answered by maintaining a grid chart of original documents against reports in circulation, showing which original documents are copied in which reports.

Question

1   What document number should be used for filing the copy control grid chart?   (3 min)

ASSIGNMENT

Concluding the mail-order wine club assignments (see end of chapter 15), revise your system design to take account of the following changes.

1) The club decides to branch out by selling wine accessories (corkscrews, decanters, etc.) and books about wine. The accessories are to be despatched by the club, accompanying any wine order made at the same time, but the despatch of the books is to be sub-contracted to a bookseller. This means that the book order-lines and delivery address, etc. must be passed to the bookseller, who is also to be paid the book price less 20% discount.

There will be a maximum of 20 accessories and 50 book titles.

2) The government announces a new three-tier plan for VAT. The regular rate is to be 15%, but books are to carry a low rate of 5%. Luxury items, including any wine having an alcohol content greater than $8\frac{1}{2}$%, are to be taxed at 30%. One wine on the present list has the low alcohol content.

3) A long strike by Post Office staff is threatened, and this would cripple the club's business. An emergency system is to be catered for, whereby replacement order forms will be delivered with the wine. Members are to be supplied with credit transfer slips, which they can present to their own banks, by means of which they can remit the total of their order. The front of the slip must conform to the standard inter-bank credit transfers, but the back of the slip can be designed to contain details of the orders. The club's own bank will send them the slips after they have been cleared through the banking system.

ANSWER POINTERS

Section 16.1

1   Curtail the conversion period by a one-shot conversion of the outstanding balance when the volume has been reduced to a manageable level. Provide interactive facilities for checking master record existence.

2   Job aids such as: procedure manuals, user guides and handbooks; wall charts and notices; booklets of codes, tables, menus; workstation equipment (e.g. calculators, telephones, waste-bins, filing equipment, furniture); input and output forms and documents; training course with objectives, instructor

and evaluation procedure. (It would be the same list for a batch system - that was a red herring.)

It should be remembered that training of users may be required not just to ensure correctness of input and output procedures, but to ensure that full advantage is taken of the opportunities and facilities presented by the system. The management may wish to combine the message system training with other training which will develop the effectiveness of staff.

3   The best man or men for the job. The prime candidates for taking overall responsibility, planning and coordinating the training and giving the instruction, including the preparation of job aids, are the user represent-atives on the project team, and/or other user managers and supervisors. If they don't know enough to do this job, what has gone wrong?

4   For computer operation: operating manuals, operating instructions (run sheets) and computer run schedules to local computer operations standards. Preparation of these is often the responsibility of programmers and oper-ations personnel.

If there is data control or data preparation within the data processing department, data control and data preparation instructions or manuals are needed. These could be prepared by data control or data preparation supervisors.

For system maintenance: the system, program and testing documentation prepared in accordance with standards. If the documentation philosophy described in this book is followed, together with a good approach to program and testing documentation, there will be little further documentation required. It will be mainly a question of tidying up what documentation there is.

Section 16.2

1   Carefully preserve the test cases submitted by programmers, analysts and users (for example, in test files on backing storage) and the document-ation of the tests and their results. When a regression test is needed, re-run the test cases and compare the new results with those previously saved.

The collected test cases comprise an 'audit test pack' which can be used at any point to help demonstrate the continued proper functioning of the system.

Section 16.3

1   Although the project budget and timetable were agreed at the outset, it is just prior to implementation that nearly all the budget has been spent and there is little to show for it in an operational sense. The sponsor is anxious to get something for his money. These anxieties are likely to be heightened if users have not been much involved in system development but are 'waiting for the data processing department to deliver'.

Another possibility is that slippage has occurred on earlier target dates. Testing and acceptance testing, which provide little further advance which will be considered tangible, are easy candidates for skimping to get back on schedule.

Sometimes it is right for the analyst to be hard-nosed about testing, even when it isn't what the sponsor wants to hear.

2   Handover: supplying a data processing system to perform some of the organisation's functions.

Cutover: using the newly supplied system in preference to the old one.

Section 16.4

1   9.1.

Appendix A Mathematical models

LEST SQUARES LINE

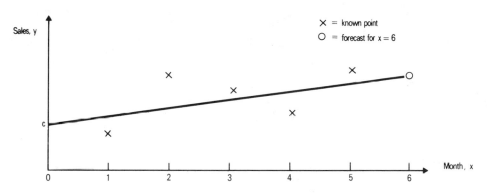

Fig. A.1: Graph of sales per month

This forecast fits a straight line to the known points in such a way as to minimise the sum of the squares of the distances of the known points from the line. If the equation of the line is $y = mx + c$, where m is the slope of the line and c is its point of intersection on the y axis, then it can be shown that with n known points

$$c = \frac{(\sum x^2)(\sum y) - (\sum xy)(\sum x)}{n\sum x^2 - (\sum x)^2}$$

$$m = \frac{n\sum xy - (\sum y)(\sum x)}{n\sum x^2 - (\sum x)^2}$$

EXPONENTIAL SMOOTHING

This is a weighted moving average which is easily computed and permits fine tuning to allow the forecaster to try to strike a balance between taking account of genuine trends but not over-reacting to the latest figure. The balance is struck by means of a constant, usually called **alpha**, which determines how much weight the latest observation is to enjoy; the balance of the weight, 1 – **alpha**, is applied to the weighted moving average that was calculated on the previous occasion. Thus

new forecast = ((1 – **alpha**) x old forecast) + (**alpha** x latest observation)

If **alpha** is set high, say 0.5, the old evidence is quickly discarded and the forecast reacts strongly to the latest evidence. If **alpha** is set low, say 0.1, then the past evidence predominates in the forecast. There is a problem of initialising the 'old forecast' when exponential smoothing is first applied; if no other evidence is available, the old forecast may be set equal to the first observation.

The least squares line tends to extrapolate the past trend; if **n** is large, it will be slow to respond to changes in the trend, if **n** is small it will be quick to respond to changes in the trend. Exponential smoothing tends to lag behind the trend; if **alpha** is set low, it will be slow to respond to the trend; if **alpha** is set large, it will be quick to respond to the trend.

STOCK CONTROL MODEL

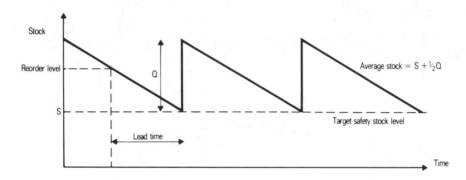

Fig. A.2: How the replenishment cycle affects stock over time

The average stock will be $S + \frac{1}{2}Q$ where S is the desired safety or 'buffer' stock and Q is the quantity of stock reordered at a time. The quantity S can be set arbitrarily or, if the variance of demand in the lead time is known, S can be set to give a known probability of run-out such that the cost of holding an additional unit of safety stock is equal to the expected costs of run-out. The reorder level is S plus the average demand in the lead time.

If the cost of reordering is R per reorder, then for an annual demand of D units (resulting in D/Q reorders being placed) there will be total annual reordering costs of $(D/Q)R$. If the cost of stockholding is I per unit per annum, then the annual stockholding cost will be $I(S + \frac{1}{2}Q)$. So the total annual cost, C, is

$$C = SI + \frac{1}{2}QI + (D/Q)R$$

Setting $\frac{dC}{dQ}$ to zero to minimise C gives

$\frac{1}{2}I - DR/Q^2 = 0$, and the optimum reorder quantity $Q = \sqrt{2DR/I}$

This model is also applicable in determining the optimum length of a production run. The quantity R is in this case the one-time cost of setting up the run and Q is the number of units to be produced. (In both cases, it is assumed that demand in the lead time is regular. The answers are not particularly sensitive to a demand which has small fluctuations; but if demand is widely variable, it may be preferable to find Q by simulation.)

LINEAR PROGRAMMING

Suppose two goods, widgets and gadgets, are made by casting and grinding. We have 200 hours of casting machinery time available and 300 hours of grinding machinery time available. A ton of widgets uses 30 hours of casting time and 20 hours of grinding time, while a ton of gadgets uses 20 hours of casting time and 40 hours of grinding time. Widgets bring profits of £6000 per ton and gadgets £5000 per ton. We can sell all we can produce and we wish to maximise profit.

So, the total profit will be 6000**w** + 5000**g** where **w** is the number of tons of widgets made and **g** is the number of tons of gadgets made. The limits on production mean that 30**w** + 20**g** must be less than 200 (i.e. total casting time for planned widget and gadget production must not exceed 200) and 20**w** + 40**g** must be less than 300 (total grinding time).

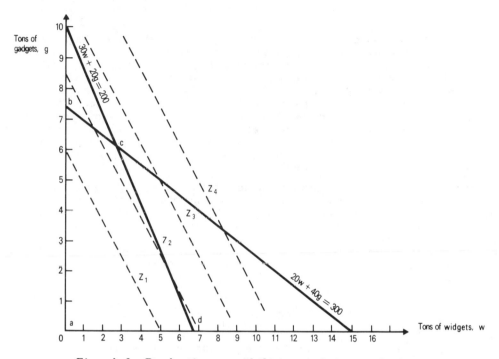

Fig. A.3: Production possibilities and iso-profit lines

The constraints are illustrated in Figure A.3. The production possibilities are constrained to **w**, **g** combinations that lie within the region **abcd**. The parallel dashed iso-profit lines join all the points that give equal contribution to profit. So the line Z_1 shows that 6 tons of gadgets and 0 tons of widgets gives as much profit as 0 tons of gadgets and 5 tons of widgets. It charts all the intermediate combinations that give the same profit. Profit increases with higher **Z** lines (further away from the origin), so we seek the production possibility that lies on the highest **Z** line. This is clearly at point **c**.

Although a graphical solution has been found here, the solution can also be found mathematically. Most computer manufacturers supply standard linear programming program packages for this purpose.

Appendix B   Sampling

In each of these cases, it is assumed that a representative sample has been or is to be obtained, e.g. by randomly selecting the cases. If the sample is chosen in a more convenient manner, e.g. by taking every ith record in a body of records, i should not correspond to any pattern in the data. It is assumed that 95% confidence levels will be satisfactory. If the decisions concerned are very sensitive to error in the inferences, with severe effects, the analyst should seek the help of a statistician. The precision in the example answers is unjustified except by the tutorial purpose. The constant of 1.96 can of course be rounded to 2 for computation; I have left the figure precise so that the reader can link this appendix to standard statistical texts should he desire.

1   Estimating the range of the population mean, given a large sample.

Van drivers are observed to record standing order amendments on a new form on 100 occasions; the number of observations is called **n**. The mean time ($\bar{x}$) to complete an order amendment was 5 minutes, with standard deviation (**s**) of 2 minutes.

The period 5 minutes is the best estimate that can be made of the true population mean, given this data. Allowing for sampling error, what would be reasonable figures to take as the upper and lower possibilities of the population mean?

Limits of population mean $= \bar{x} \pm 1.96(s/\sqrt{n})$
$$= 5 \pm 1.96(2/\sqrt{100}) = 5 \pm 0.392$$
$$= 4.6 \text{ to } 5.4 \text{ minutes.}$$

2   Estimating the size of sample needed for stated accuracy limits on the estimate of the population mean.

How many observations must be made of the time to record standing order amendments, if it is desired that the resulting estimate of the mean, plus or minus 10%, will embrace the true figure?

The sample size $= \dfrac{4 \times 1.96^2 \times s^2}{L^2}$

where L is the range between desired limits of accuracy. If L is retained as a percentage of the mean, naturally **s**, the standard deviation, must also be expressed as a percentage. Suppose our previous experience allows us to guess that **s** is 30% of the mean, i.e. we expect about half the observations to lie in the band mean – 30% to mean + 30%. Then the desired sample size $= 4 \times 1.96^2 \times 0.3^2 \div 0.2^2 = 35$.

If there is uncertainty over **s**, the analyst can explore the sensitivity of sample size to other values of **s**, to get a feel for his problem, or he can refine his estimate of **s** after some samples are taken.

3   Estimating the range of a population proportion from a large sample.

A total of 200 statements have been inspected and it is found that 6% have more than 10 transactions. What is the reasonable maximum proportion of statements with more than 10 transactions, in the whole population of statements?

The limits are $p \pm 1.96\sqrt{pq/n}$ where **p** is the proportion and $q = 1 - p$.

In our example, the population proportion range is

$$0.06 \pm 1.96\sqrt{0.06 \times 0.94/200}$$

and a reasonable upper limit for the proportion of statements with more than 10 transactions is 9.3%.

4   Estimating the sample size required for desired limits of accuracy on the estimate of the population mean.

How many orders must be inspected to establish, within plus or minus 5 percentage points, the proportion of all orders which required special approval from the credit manager?

$$n = 4 \times 1.96^2 (pq)/L^2$$

This formula requires a prior guess of the likely proportion, p, which will have to be based on experience (or estimated from a preliminary sample). Suppose p is thought to be about 20%; then, since the total range is 10 percentage points,

$$n = 4 \times 1.96^2 (0.2 \times 0.8)/0.1^2 = 246.$$

Appendix C Further assignments

SPEEDFERRIES LTD

Assignments

This case can be used either for practice in the design and implementation of a complete system or for smaller assignments, which allow practice of specific skills, using the context of the case as background. If implementing the complete system on, say, a microcomputer plainly too small to permit realism, the approach could be to build a scale model of the full-size system by stringently limiting the file sizes (perhaps even to the point where they can all be held in memory). An alternative assignment with a microcomputer could be to prepare it for use as an intelligent terminal in the system.

Examples of the smaller assignments mentioned are as follows.

1   Draw an entity-relationship diagram for the booking system. Define the attributes and primary keys of the entity and relationship relations you envisage.
2   Design the dialogue and VDU layouts for Booking Enquiries and Confirmations.
3   Design the top copy Confirmation and Invoice (see additional notes at the end of this appendix).
4   Write a program specification for a program module which is to establish whether or not a booking can be accepted. The module is to receive a ·booking enquiry as a parameter (this may be assumed to have been validated as far as possible, short of access already having been made to the database), and return a Yes/No indicator showing acceptability. The relevant file and record specifications should be included. Invalid enquiries should be returned with an error code.
5   Estimate the online backing-storage requirements (for data files) of the centralised computer. Estimate the message traffic to be handled by the centralised computer (a) assuming dumb terminals, (b) assuming terminals can be programmed locally to validate and edit messages sent to or from the centralised computer and (c) assuming terminals can locally validate and edit and have local intelligence sufficient to store all the details of one booking enquiry and confirmation, **and** drive the slow line printer to produce the Confirmation and Invoice.

Additional assumptions made in completing assignments should be specified, and documentation should conform to standards.

The background

Speedferries Ltd is a drive-on, drive-off cross-channel car ferry operating over the following routes: Folkestone-Boulogne; Felixstowe-Zeebrugge; Harwich-Hook; Southampton-Cherbourg. Nine ferry-boats are presently in service, three on the Folkestone-Boulogne route, two each on the others. A schedule of crossings (routes, dates, times, boats) is worked out for a period 12 months ahead and bookings are accepted on these crossings. There are an average of 6 return crossings per day on the Folkestone-Boulogne route, and two per day on the other routes. The boats carry about 1000 passengers and about 1000 metres total length of vehicles – cars, lorries, coaches (possibly with caravans or trailers). However, no two boats have exactly the same capacities.

Advance bookings are made by passengers directly with the company's offices at 39 Wapping Steps, London E2A ZAA, or with the company's local offices in the English ports, or with travel agents. Travel agents may deal with Wapping Steps, or with the local office. 60% of all bookings are made at Wapping Steps. 50% of bookings are made by passengers direct, without an agent. Of agency bookings, about 95% of them are made by 100 or so agencies who are regular customers; a very large number of minor agencies are involved in the other 5% of bookings.

About 30% of bookings are made by telephone to Wapping Steps. The client has his booking confirmed orally and is given a booking reference number to quote should he have a query. A follow-up printed Confirmation and Invoice is despatched as soon as possible afterwards – these are all sent from Wapping Steps. 60% of bookings are made by mail to Wapping Steps on a booking form accompanied by a cheque. When the cash is received, a ticket is despatched as soon as possible. 10% of bookings are made by personal callers at the local offices or at Wapping Steps; the callers pay cash and are given a ticket forthwith. 10% of the telephone bookers also call at a local office to pay their invoices in cash and receive tickets.

A booking is always for at least one adult passenger who (even if he is travelling alone) is known as the 'group leader' and whose name and address appears on the invoice and ticket. A booking may be for a group of adults, children and infants. Children travel at reduced fare and infants travel free, but the number of infants in a group is still recorded on the ticket. 10% of bookings are for foot passengers or passengers with bicycles or solo motor cycles. A charge is made for a bicycle or a solo motor cycle, but no space is reserved on the vessel.

A maximum of one vehicle may be involved in a booking. It is charged at tariff A (under 4.00 metres length), B (4.00-4.49 metres), C (4.50-4.99 metres), D (5.00-5.50 metres) or E (as tariff D plus a certain charge per metre or part metre over 5.50. The overall length of a vehicle and its trailer or caravan is taken. A motor cycle with a sidecar is treated as a vehicle. The vehicle registration number and driver's name are recorded on the ticket.

The group leader may book one or more cabins for the crossing on any route except Folkestone-Boulogne. There are seven types of cabin available: bridge deck cabin, 2-berth special cabin, 2-berth standard cabin, 4-berth standard cabin, 4-couchette cabin, reclining seat, and day cabin. The cabin accommodation of each boat is different, but a standard cabin number is used on all boats. This is generally of the form **nmm** where **n** is a single digit 1-7 showing cabin type and **mm** is the cabin serial number, starting at 01 for each cabin type (the highest serial number within a type is equal to the number of cabins of that type on the boat). In the case of cabin type 6 (reclining seat), a three-digit serial number (**mmm**) is used. Past bookings have been distributed: 40% no cabin, 50% one cabin, 5% two cabins, 3% three cabins, 1% four cabins, 1% five or more cabins. The cabin number (and type) is shown on the ticket. An extra charge is made for all cabins.

Most bookings are for specific crossings, but some 5% are left 'open subject to space' on a designated route – these are mainly the inward part of an outward/inward return booking. Cabins cannot be booked on an open-dated crossing. In 80% of bookings, the 'outward' journey originates from an English port. Bookings made in other countries are handled mainly through travel agents in those countries. In 80% of cases, a return crossing is booked. A return booking for a vehicle always involves the same vehicle outward and inward, but in 15% of return bookings the number of passengers differs on the outward and inward legs. Only 85% of the possible passenger and vehicle space is available for pre-booking, to allow for casual customers.

The plan

It is planned to provide booking clerks at Wapping Steps and local offices with work stations comprising an alphanumeric visual display unit and a slow printer. There may be more than one VDU to a printer. The clerk is able to make a Booking Enquiry over the VDU and learn whether or not the booking can be accepted. If it can, he may confirm it, whereupon a combined Confirmation and Invoice (top copy) and Ticket (second copy) is to be printed forthwith on the printer. For telephone bookings, the confirmation and Invoice will be despatched to the customer in a window envelope and the ticket will be filed to await remittance. In other cases, the amount paid is also entered and both parts are given or sent to the customer. If a customer miscalculates his remittance, there may be a balance due, in which case the Ticket is to be withheld pending receipt of the balance.

  The VDUs and printers are to be connected to a large centralised computer at Wapping Steps. The amount of local intelligence in the VDUs has not yet been decided.

  It is also intended that a manifest, listing all the bookings by group leader within vehicle type, and a cabin report, listing cabin bookings by group leader within cabin number, will be produced locally in the evening before the day of sailing. At head office, a monthly report is to be prepared summarising the forward loading of all scheduled crossings. Accounts of travel agents' indebtedness, after allowing for their commissions, are also to be maintained, with statements sent monthly. A VAT list showing net invoice amount and VAT amount, ordered by invoice number within invoice date, is also to be prepared monthly.

Confirmation and Invoice – additional notes

The top copy details are copied through to the Ticket. Tickets may also be posted in window envelopes. They are cancelled with a stamp as the passenger boards, but not collected.

  The Confirmation and Invoice is to bear the following pre-printed messages.

'We are pleased to confirm your booking. Please check and ensure that all the details shown are correct. Your remittance in full is required not later than 30 days before the outward sailing date (or within seven days of the date of this invoice, whichever is the later). If payment is not received within this period, we reserve the right to cancel the complete reservation without notifying you.'
'Please return this invoice with your remittance to help us despatch your ticket speedily.'
'Telephone (01) 499 9999'
'Telex SPEEDY LONDON'
Latest reporting time: 35 minutes before sailing, coach parties 60 minutes before sailing.'
'Cancellations: tickets must be returned before any refund can be made.'
'Important: see conditions of carriage which govern all passengers and their property.'
'Registered Office: Speedy Tower, Park Lane, London W1Z AXX.'

Appendix D The BCS Code of Practice

INTRODUCTION

This Code of Practice (reproduced with the permission of the Society) is directed to all members of the British Computer Society. It consists, essentially, of statements which prescribe the minimum standards of practice, to be observed by all members.

The code is concerned with professional responsibility. All members have responsibilities: to clients, to users, to the State and society at large. Those members who are employees also have responsibility to their employers and employers' customers and, often, to a Trade Union. In the event of an apparent clash in responsibilities, obligations or prescribed practice the Society's Secretary-General should be consulted at the earliest opportunity. The Code is to be viewed as whole: individual parts are not intended to be used in isolation to justify errors of omission or commission.

The Code is intended to be observed in the spirit and not merely to the word. The BCS membership covers all occupations relevant to the use of computers and it is not possible to define the code in terms directly relevant to each individual member. For this reason the Code is set out in two levels to enable every member to reach appropriate interpretations.

Level One: a series of brief statements which define the elements of practice to be observed.
Level Two: the rationale for the Level One statements.

Level Two is not intended as guidance on **how** to carry out the Code of Practice, but only to provide an explanation of its meaning and the reason for including the statement at Level One. Where examples are given of how to apply the code, these are simply to clarify the meaning. Many of the clauses may seem to state the obvious, but much that goes wrong in computer use does so because the obvious has been overlooked.
Terminology   The following conventions apply to the reading of this code.
1 'He' (etc.) includes 'she' (etc.).
2 'Client' is any person, department or organisation for whom the member works, or undertakes to provide computer-based aid, in any way.
3 'User' is any person, department or organisation served by computer-based systems.
4 'System' means all applications involving the use of computers. The term does not imply any particular mode of processing (e.g. dedicated, batch or transaction). 'System' may be interpreted as encompassing noncomputer procedures such as clerical, manual, communication and electromechanical processes.
Layout   Level One appears on the left, Level Two on the right, the rationale being opposite the appropriate statement.
Further reading   The BCS Librarian maintains a list of publications suggested for each of the seven main clauses of this Code.

LEVEL ONE

In the practice of his profession the member will, to the extent that he is responsible:

1   Personal requirements

1.1   Keep himself, and subordinates, informed of such new technologies, practices, legal requirements and standards as are relevant to his duties.

1.2   Ensure subordinates are trained in order to be effective in their duties and to qualify for increased responsibilities.

1.3   Accept only such work as he believes he is competent to perform and not hesitate to obtain additional expertise from appropriately qualified individuals where advisable.

1.4   Actively seek opportunities for increasing efficiency and effectiveness to the benefit of the user and of the ultimate recipient.

2   Organisation and management

2.1   Plan, establish and review objectives, tasks and organisational structures for himself, and subordinates, to help meet overall objectives.

2.2   Ensure that any specific tasks are assigned to identified individuals according to their known ability and competence.

2.3   Establish and maintain channels of communication from and to seniors, equals and subordinates.

LEVEL TWO

1.1 Others will expect you to provide special skills and advice and, in order to do so, you must keep yourself up-to-date. This is true for members of all professions, but particularly so in computing which is developing and changing rapidly. You must also encourage your staff and colleagues to do the same, for it is impossible to retain your professional standing by relying only on the state of your knowledge and competence at the time you achieved professional status.

1.2 Take action to ensure that your hard-won knowledge and experience are passed on in such a way that those who receive them not only improve their own effectiveness in their present positions but also become keen to advance their careers and take on additional responsibilities.

1.3 You should always be aware of your own limitations and not knowingly imply that you have competence you do not possess. This is of course distinct from accepting a task the successful completion of which requires expertise additional to your own. This point is central to the BCS Code of Conduct; you cannot possibly be knowledgeable on all facts, but you should be able to recognise when you need additional expertise and information, and where to find it.

1.4 Whatever the precise terms of your brief, you should always be aware of the environment surrounding it and not work solely towards the completion of the defined task and no more. You must regard it as part of your duty to make your client aware of other needs that emerge, unsatisfactory procedures that need modification and benefits that might be achieved. You, as an innovator, should take into account the relevance of new methods and should always be looking for the possibility of additional benefits not foreseen when the project was planned. You must also look beyond the immediate requirements to the needs of the ultimate user. For example: the invoice your system produces may be right for company accounting procedures but confusing for the person who is being asked to pay against it.

(This section of the Code is concerned with broad principles. Management of development projects is covered in detail in Sections 5 and 6; management of operational projects in Section 7. Since computer management is still management, the normal principles applicable to any kind of management apply here also.)

2.1 It is dangerously easy for you as a computer professional to become fully engrossed in the problem of the moment, and to lose sight of the overall objectives of the organisation. Computing, no less than any other discipline, is an organic component of the organisation, and you should continuously ensure that the path you are following is in line with the objectives of that organisation. You must make use of the well established management practices of monitoring and review to ensure the area of work for which you are responsible is making its maximum contribution.

2.2 When delegating work to your subordinates ensure that as far as possible the tasks will develop their competence and increase their motiv-ation. However, you must also ensure that the principles implied in 1.3 are observed or you will be faced with a dissatisfied user who is not receiving the service to which he is entitled.

2.3 It is often assumed that communication will look after itself, but good communication is vital to business success. You must ensure that formal channels of communication exist upwards, downwards and sideways in the organisation for which you are responsible. It is difficult to over emphasise this point in connection with computer work which by its nature requires

LEVEL ONE

2.4 Be accountable for the quality, timeliness and use of resources in the work for which he is responsible.

3   Contracting

3.1 Seek expert advice in the preparation of any formal contract.

3.2 Ensure that all requirements and the precise responsibilities of all parties are adequately covered in any contract or tendering procedures.

4   Privacy, security and integrity

4.1 Ascertain and evaluate all potential risks in a particular project with regard to the cost, effectiveness and practicability of proposed levels of security.

4.2 Recommend appropriate levels of security, commensurate with the anticipated risks, and appropriate to the needs of the client.

LEVEL TWO

constant interaction between the members of the computer organisation and, most importantly, with the user. Furthermore, you will find that communication skills can be improved considerably by formal training and this should be included in your training plans as a high priority item.

2.4 High on the list of your professional duties will be the requirement to provide a service of agreed quality, on time and within budget. Beyond that, of course, is the requirement for contingency planning and the need to make others affected aware of difficulties and dangers if these are foreseeable. For this you, as a professional, are responsible. You cannot turn your back on a problem once recognised, and hope someone else will solve it or that it will simply go away. Action taken to minimise the impact of such problems will, in the end, ensure a smoother running organisation.

(This section is included in the Code because some formal agreement – even if not a specific contract – is needed before any project is started. Commitment and definition of responsibilities are essential, in advance of action.)

3.1 In the same way as you would expect to be consulted in your field as a computer professional, be ready to consult other specialists when you require assistance or guidance in drawing up contracts or in matters such as commerce, finance, tax, legal or risk evaluation. Much of your time can be saved in this way, to say nothing of avoiding the potential dangers of a badly drawn up contract or wrong assessment of a legal situation. Many of these areas have become defined as standard practice and a number of professional bodies provide 'standard contract' forms as a guide to their own members which help considerably to reduce problem areas.

3.2 In the same way as you would carefully review the completeness of the detail for a systems specification, it is necessary to review the totality of the detail to be covered by a contract. Take care to ensure such items as provision of accommodation, typing, data preparation, responsibility for media security and standby arrangements are not forgotten. Apart from the problems which will arise if these things have been overlooked, the profitability of your contract will be adversely affected. Again, communication enters into this and you will need to ensure that everyone who is party to the contract is fully aware of his obligations under the contract.

(A system is at risk from the moment that the project which develops it is first conceived. The risk remains until after the system is finally discontinued, perhaps indefinitely. Threats to security range from incompetence, accident and carelessness to deliberate theft, fraud, espionage or malicious attack.)

4.1 One of your more difficult responsibilities is that of determining the value of a system in terms of what would be lost if system security were to be breached (e.g. damage to national security by leaks of military data, personal privacy by leaks from medical records or fraud by access to financial information.) However, a view is required to aid decision making, covering how much should be spent on system security in at least these four areas:

protection,   preventing threats from becoming reality;
detection,   in time to take suppressive action;
suppression,   to limit the effect;
recovery,   to rectify and get the system going again.

4.2 You still need to remember that you must give attention to some areas of risk which are mandatory such as those covered by legislation for health and safety at work. However, risks exist in connection with the security of your hardware, software, data systems and people, all of which should be identified and recommendations made.

LEVEL ONE

4.3 Apply, monitor and report upon the effectiveness of the agreed levels of security.

4.4 Ensure that all staff are trained to take effective action to protect life, data and equipment (in that order) in the event of disaster.

4.5 Take all reasonable measures to protect confidential information from inadvertant or deliberate improper access or use.

4.6 Ensure that competent people are assigned to be responsible for the accuracy and integrity of data in each data file and each part of an organisation's data base.

5   Development

5.1 Exercise impartiality when evaluating the implications of each project with respect to its technical, moral and economic benefits.

5.2 Effectively plan, monitor, adjust and report on all development, acquisition or replacement projects.

5.3 Ensure that effective standard procedures and documentation are available and used.

5.4 Specify the system objectives, completion date, cost and security requirements with the client and the necessary criteria for their achievement.

5.5 Ensure that the client can participate in all stages of problem analysis, system development and implementation.

LEVEL TWO

4.3   Situations are always changing and people are liable to become lax in observing routine practices. You will therefore find an on-going security audit extremely valuable in keeping people aware of security requirements and procedures, and in the identification of weaknesses and loopholes in the security system. Moreover security arrangements should be reviewed periodically in the light of developing technology and new methods of breaching security.

4.4   Naturally, the safety of people is your first priority. The data is the next priority, and proper backup facilities for re-creation of data files should exist. Equipment should be replaceable and normally insured.

Your staff should be trained to react with regard to these priorities. Data processing centres are potentially vulnerable to deliberate damage with consequential seriousness to the business of the organisations involved. As a professional you will be concerned to treat security drill as a serious matter and to carry it out regularly along with training all involved to be on the lookout for anything unusual.

4.5   Your responsibility for confidentiality of information is at least as great as that of members of other professions. The job is even more complex by reason of the speed, capacity and facility for data exchange by computers. Frequently, personal information will be under your control, and you should always be aware of the spirit and letter of relevant legislation written to protect the individual.

4.6   You must take direct action to give responsibility to specific individuals to ensure the accuracy and integrity of data within each system. Whilst this is important for any system, however simple, it becomes even more significant in more complex data base and communications environments.

('Development' in this context means not only all the work involved in order to reach the stage where a viable computer system is ready to become operational, it also includes maintenance and enhancement work.)

5.1   Your responsibility in a project will give you opportunities to make decisions based on your personal views and preferences. The line between personal bias and professional opinion becomes somewhat finely drawn. To avoid finding yourself on the wrong side of the line, always make sure you are aware of your client's objectives and the benefits he is looking for, and be careful not to lose objectivity through enthusiasm created by the latest development of technology.

5.2   This principle is no different from that applying to many other professionals in other fields, but your attention is drawn to it as it is essential to business control in any organisation.

5.3   A characteristic of professionals is that they depend on the operation of a series of standards and procedures for efficiency and effectiveness. This is no less true of the computer professional. You should ensure that the standards you lay down do not cause inhibiting rigidity, but provide a framework within which individuals know how the work is to be done, when and by whom.

5.4   Always ensure you produce a clear statement with qualified objectives wherever possible which can be agreed with the client. It is all too easy to overlook this point in the general rush of business life: when committing agreements to paper it is frequently a neglected activity. For large projects covering a significant span of time, objectives should in fact be reviewed to ensure that the project is still relevant in the light of changing circumstances.

5.5   The systems you develop belong to the client, and therefore he needs to maintain overall control and be given opportunities to exercise it. Therefore you should seek his involvement in key project activities, e.g. the specification, quality control and provision of test data. You should

LEVEL ONE

5.6   Ensure that each task is completed to a defined level before the next dependent task is started.

5.7   Specify and conduct program tests and system tests to ensure that all system objectives are met to the satisfaction of the client.

5.8   Ensure that systems are designed and sufficiently documented to facilitate subsequent audit, maintenance and enhancement, and accurate comprehension by users.

5.9   Ensure that input and output are designed to obviate misunderstanding.

5.10 Ensure that there are adequate procedures available to delete erroneous, redundant and out of date data from files.

5.11 Ensure that adequate procedures are available which will, with the minimum of inconvenience, restore data files and program files to their required conditions in the event of data loss, corruption or system failure.

5.12 Ensure that projects are completed with technical soundness, using the most appropriate technology and within time and cost constraints.

6   Implementation

6.1   Ensure that adequate provision is made for user and operations staff training in all functions of the system for which they are responsible.

LEVEL TWO

encourage and help the client to achieve the right level of involvement not least because in this way you ensure you produce the system that the client requires.

5.6 A task may be anything from specifying a system to determining the size of a piece of detailed code. While many tasks will be executed in parallel, dependent tasks should be completed sequentially with nondependent activities within them overlapped. But you should not, for example, start writing a program in advance of a complete specification if you wish to avoid duplication or waste of effort in reprogramming.

5.7 It is clearly necessary for you to plan and test each program separately and then all programs together as a complete suite, followed by the computer elements together with the rest of the system. The objective is to prove the system functions as intended and not merely to detect errors. Remember the client should be involved in the testing to achieve the objectives in 5.5.

5.8 It is essential, at the original design stage, that you consider and provide for the needs of future audit and of modification. Documentation should clearly indicate where the audit trail lies. Documentation should also assist trouble-shooting and enable modification to be undertaken with minimal reprogramming and the smallest possible impact on operations. Also your users will require documentation in a convenient form to ensure the proper use and exploitation of the system.

5.9 The input and output of a system are normally prepared or received by nontechnical users and consequently must be designed to simplify business life rather than add extra burdens. Input and output should be readable – avoid jargon, unfamiliar codes and abbreviations – and provide clear headings and such things as page numbers. Moreover, whenever possible, the power of the computer should be used so as to permit the maximum use of plain English.

5.10 It is part of a sound approach to consider not only the immediate use of a system but also its effectiveness during a life which will be as long as it continues to meet its objectives. During this life, redundant data is bound to accumulate and it will be essential to have procedures for clearing it out. Without proper procedures, undisciplined corrections or deletions may occur, thereby compounding the problem (see also below).

5.11 This is complementary to 4.1. The design stage is the time to ensure that the restorative procedures are incorporated.

When an operational disaster occurs it will be too late to start thinking about such procedures.

The emphasis in 5.10 and 5.11 is on clear procedures to protect data and programs from corruption rather than relying on ad hoc correction by individuals who may subsequently be the only ones who know what they have done.

5.12 Cost and service are criteria of an effective system rather than technical ingenuity. The technology to be exploited should be the best for the purpose in view, not necessarily the latest or the most sophisticated.

(The term is used here to describe the transition from development to full operation.)

6.1 You should not consider the task complete when you have seen the new system through to implementation. Your professional duty requires you to see that the system can be used effectively by your client's staff.

Each new system will bring with it, to some degree, new approaches, new techniques and new ways of doing things. These have to be explained to your users who may show resistance to change because of their previous experience. You should recognise that they will require time to become familiar with the new system and to gain confidence both in the system and

LEVEL ONE

6.2 Effectively plan, monitor, adjust and report upon all activities concerned with the changeover from development to operational running.

6.3 Ensure expeditious and economic completion of implementation consistent with adequate testing and security.

7 Live systems

7.1 Plan and operate efficient and reliable processing services within defined budgets.

7.2 Monitor performance and quality and arrange regular reviews of the efficiency, effectiveness and security of live systems.

7.3 Plan, from the start of a project, to provide adequate maintenance and enhancement support to live systems so that they continue to meet all requirements.

7.4 Establish good liaison with users and provide proper facilities for dealing with enquiries and day-to-day problems concerning the use of systems.

Adopted by Council
26 July 1978

LEVEL TWO

their own ability to meet the new conditions. Training in advance of implementation and follow up sessions aimed at increasing understanding and seeking the user's view on improving operation will do much towards gaining acceptance of the new system.

6.2 These are vital parts of the design and development process. Your plans and schedules must be accurate and complete in detail for all resources involved whether user's or DP. Furthermore constant reviewing will be necessary as implementation draws near. Your communication responsibilities will be put to the test here, as all who are affected will need to be advised of changes and be given the opportunity to comment. Again, the opportunity presents itself to help generate the understanding, confidence and sense of involvement so necessary to successful implementation and subsequent operation.

If you fail to carry out these functions effectively, operational dates will be jeopardised and, almost certainly, implementation costs will be higher than they need be.

6.3 Here you are involved in a professional judgement or trade off between reasonable costs and acceptable levels of proving or testing. If you cut corners by, say, reducing system testing time, then the likely effect on the operation elsewhere should be evaluated and made known to those who should know.

(This section is concerned with the ongoing operation of systems handed over by design and development staff.)

7.1 'Processing services' covers all the activities between reception of data and delivery of results.

You must ensure that these services are provided efficiently to users who are just as dependent on these as they are on the application for the well being of their business.

7.2 The dynamic nature of most business environments means that over a period a system may gradually provide the user with a service inferior to that originally planned. Your post-implementation reviews will be all the more effective if you check not only how well the system is meeting its original objectives, but also the continuing validity of the original objectives in the light of current business requirements.

7.3 Much of the criticism computer applications receive is traceable to their failure to respond, by means of modification, to changing conditions. Either modifications do not happen, or they are implemented haphazardly over too long a period. If you ensure that your project plans include provision of a formal system to control the enhancement of programs, and identify the need for appropriate maintenance resources, you will avoid user dissatisfaction arising from this type of problem.

7.4 One of the most important areas where your professionalism will be tested will be in maintaining continuous formal and informal liaison with your users. Everyone concerned with the services you are responsible for should know and understand the need for formal channels of communication. In particular, do not forget to ensure that these exist to cover the special circumstances which arise in emergencies.

Appendix E The BCS Code of Conduct

A professional member of the BCS

1 Will behave at all times with integrity. He will not knowingly lay claims to a level of competence that he does not possess and he will at all times exercise competence at least to the level he claims.

2 Will act with complete discretion when entrusted with confidential information.

3 Will act with strict impartiality when purporting to give independent advice and must disclose any relevant interest.

4 Will accept full responsibility for any work which he undertakes and will construct and deliver that which he purports to deliver.

5 Will not seek personal advantage to the detriment of the Society.

Five principles make up the BCS Code of Conduct and each professional member of the society, as a condition of membership, undertakes to adhere to these principles. The principles, set out separately above, are clear but have an inevitable appearance of generality and below each principle is supported by a number of notes for guidance which will help in specific interpretation. Members of the Society will readily appreciate that continued evidence of their determination to abide by the Code will ensure the public trust and confidence in computer professionals which is so necessary to the continuing effective use of computers.

The Society, through its Professional Panel, is ready at all times to give guidance in the application of the Code of Conduct. In cases where resolution of difficulties is not possible informally, the Society will invoke the disciplinary procedures defined in its Articles of Association. These procedures involve initial discussion to establish the background for a formal complaint, the appointment of an Investigation Committee and, if the latter finds a case to answer, a Disciplinary Committee. The Disciplinary Committee is empowered to exclude from the Society; to suspend from membership for a given period; to reprimand; to admonish, or, of course, to dismiss the case.

NOTES FOR GUIDANCE

Integrity

Integrity implies wholeness, soundness, completeness: anything he does should be done competently. Where necessary he should obtain additional guidance or expertise from properly qualified advisers.

While claims to competence should not be made lightly, a member will not shelter behind this principle to avoid being helpful and cooperative; any guidance or advice that he can provide from his experience should be readily given.

He should act in a manner based on trust and good faith towards his clients or employers and towards others with whom his work is connected.

He should express an opinion on a subject in his field only when it is founded on an adequate knowledge and honest conviction, and will properly qualify himself when expressing an opinion outside his professional competence.

He should not deliberately make false or exaggerated statements as to the state of affairs existing or expected regarding any aspect of the construction or use of computers.

He should do his best to keep himself aware of relevant developments in his technology.

He should comply with the BCS Code of Practice and any other codes that are applicable and ensure that his clients are aware of the significance to their work.

Confidentiality

He should not disclose, or permit to be disclosed, or use to his own advantage, any confidential information relating to the affairs of his present or previous employers or customers without their prior permission. This principle covers the need to protect confidential data.

Many kinds of information can be considered by a client or employer to be confidential. Even the fact that a project exists may be sensitive. Business plans, trade secrets, personal information are all examples of confidential data.

Training is required for all staff on measures to ensure confidentiality, to guard against the possibility of a third party intentionally or inadvertantly misusing data and to be watchful for leaks of confidentiality arising from careless use of data or indiscretions.

Impartiality

This principle is primarily directed to the case where the member or his relatives and friends may make a private profit if the client or employer follows his advice. Any such interest should be disclosed in advance.

A second interpretation is where there is no immediate personal profit but the future business or scope of influence of his department depends on a certain solution being accepted. Whereas a salesman is assumed to have a bias towards his own company, an internal consultant should always consider the welfare of the organisation as a whole and not just the increased application of computers.

Responsibility

Trust and responsibility are at the heart of professionalism. A member should seek out responsibility and discharge it with integrity. He should complete the work he accepts on time and within budget. If he cannot achieve what he promised then he must alert the client or employer at the earliest possible time so that corrective action can be taken.

He should have regard to the effect of computer based systems, in so far as these are known to him, on the basic human rights of individuals, whether within the organisation, its customers or suppliers, or among the general public.

Subject to the confidential relationship between himself and his customer, he is expected to transmit the benefit of information which he acquires during the practice of his profession, as a result of his technical knowledge, to illuminate any situation which may harm or seriously affect a third party.

He should combat ignorance about his technology wherever he finds it and in particular in those areas where application of his technology appears to have dubious social merit.

Relationship to the Society

It is necessary to write this principle into the Code of Conduct to prevent misuse of the considerable influence that a professional society can have. Nevertheless, its impact is largely internal and the points that have been made should be read in that light.

He should not bring the Society into disrepute by personal behaviour or acts when acknowledged or known to be a representative of the Society.

He should not misrepresent the views of the Society nor represent that the views of a segment or group of the Society constitutes the view of the Society as a whole.

When acting or speaking on behalf of the Society he should, if faced with a conflict of interest, declare his position. He should not serve his own pecuniary interests or those of the company which normally employs him when purporting to act in an independent manner as representative of the Society, save as permitted by the Society following a full disclosure of all the facts.

He is expected to apply the same high standard of behaviour in his social life as is demanded of him in his professional activities in so far as these interact. Confidence is at the root of the validity of the qualifications of the Society and conduct which in any way undermines that conduct (e.g. a gross breach of a confidential relationship) is of deep concern to the Society.

He should conduct himself with courtesy and consideration towards all with whom he comes into contact in the course of his professional work.

He should have regard to the great extent that professional and other bodies depend on voluntary effort and should consider what personal contribution he can make both to the Society and to the public generally.

THE CODE AS APPLIED TO A CONSULTANT

Advice given to a client can come from (a) outside an organisation, either for a fee or as part of a supplier's marketing effort or after-sales support; or (b) within the organisation from business analysts or systems designers working directly or indirectly for a user. Irrespective of conditions of employment, consultants are expected to give sound advice and honest opinion, and to help the client to a successful planned conclusion. The following points amplify the notes for guidance in respect of consultancy work.

He should hold himself accountable for the advice given to his client, and should ensure that all known limitations of his work are fully disclosed, documented and explained.

He should not attempt to avoid the consequences of poor advice by making the language of his report incomprehensible to the layman by the use of computer jargon.

He should ensure that his client is aware of all significant contingencies and risks which could adversely affect his plans and the scale of the costs he may incur as a result of embarking on any particular DP strategy.

During the course of the work he should bring to the client's attention, at the earliest possible time, any risk that the stated objectives may not be achievable; and if the solution lies in an extension of contract he should use his best efforts to make the necessary time available at an equitable fee.

Where it is possible that decisions may be made as a result of his efforts which could adversely affect the social security, work or career of an individual, he should ensure that his clients are aware of their responsibilities to mitigate the effects of their decisions.

He should always have regard to any factors arising during his professional

assignment which might reflect adversely on his integrity and objectivity.
  He should declare to his client, before accepting instructions, all interests
which may affect the proper performance of his functions. For example

a)   a directorship or controlling interest in any business which is in
competition with his clients;
b)   a financial interest in any goods or services recommended to his client;
c)   a personal relationship with any person in a client's employment who
might influence, or be directly affected by, his advice.

  When undertaking consultancy work, he must provide a written agreement
clearly stating the basis or amount of remuneration before undertaking the
assignment. He is expected not to structure his fees in any way so as to
offset his impartiality; examples which have in the past been regarded as
suspect include fee splitting, and many cases of payment by results.
  He should not invite any employee of a client to consider alternative
employment without the prior consent of that client. (An advertisement in
the press is not considered to be an invitation to any particular person
for the purpose of this rule.)

THE CODE AS APPLIED TO A SALESMAN

Almost everyone in computing is from time to time in the position of
salesman – either in direct contact with clients and customers, or with
members of the public (who are themselves often potential customers) or with
those who, because they are dependent on the results from computing, are
in the position of clients. Salesmen are normally direct employees of their
companies, and it is implicit that whatever they promise to a customer
should be delivered by the company. A salesman must therefore act loyally
and honestly as an employee and should declare his status as a representative of his company.
  Payment by results in the form of commission to a salesman is an accepted
business practice; but in the selling of a continuing system it is probably
desirable that some or all of the commission be tied to the proper performance of the work and the long term satisfaction of the customer. He should
act in a manner based on trust and good faith towards his customer, to
ensure that he receives lasting and profitable enjoyment of his purchase.
For example

  he should accept only such work as he believes his organisation can
    produce and deliver;
  he should ensure that any agreement with his customer is explicit,
    unambiguous and complete;
  he should obey the spirit as well as the letter of any contract and of
    the law;
  he should secure an after-sales service where appropriate, commensurate
    with the kind of product supplied and the price paid;
  he should ensure that the customer is aware of any contingencies under
    which supplementary charges may be payable and the basis of such
    charges;
  he should ensure that the customer is aware of any significant risks
    which could adversely affect his plans, and of any additional work
    or expense he will or may incur in using the service or product which
    is being sold;
  he should subcontract only to responsible practitioners and organisat-
    ions;
  he should avoid illegal 'informal' price fixing, market sharing arrange-
    ments tending to falsify the process of tendering and open competition;
  he should not be party to any practice which could lead to commercial

     or other corruption;

     he should not use products commissioned and paid for by one client for another client, without the knowledge and agreement of the original client;

     he should not denigrate the honesty or competence of a fellow professional or competitor with intent to gain unfair advantage;

     he should not maliciously or recklessly injure or attempt to injure, directly or indirectly, the professional reputation, prospects, or business of others;

     he should not exploit customer relations by using either the existence of any contract or any identifiable precis or work done in any advertising or publicity material without the permission of the client.

DISCIPLINARY PROCEDURE

All members of the Society undertake to abide by the Society's Code of Conduct. It will sometimes happen, however, that someone (member or non-member) wishes to lay a complaint against a member for infringement of the Code, and this note explains the Society's procedures.

First the complaint is laid by letter with the Secretary-General. In many cases, because of the knowledge and experience that is available to members of the Society in the several areas of computer practice, the grievance can be settled there and then avoiding the time and effort of a formal enquiry. These discussions are conducted in strict confidence.

When a more difficult problem is presented an Investigation Committee can be nominated to look into the grievance and make a case to the Disciplinary Committee. Members should be aware that formulating such a case implies a commitment by the Society of funds and resources and the decision to proceed on any particular case is a matter for decision of the Finance and General Purposes Committee, a standing committee of the Council of the Society. The Disciplinary Committee will set a date for a hearing and invite the complainant and the respondent to be heard giving due notice to both parties. Legal assistance may be retained. Sanctions which can be applied include Admonishment, Suspension and Expulsion from membership. In addition, the Disciplinary Committee has power to make public the result of its findings.

There is an appeal procedure.

The Code of Conduct is administered by a representative group, and a number of members of the Investigation Committee retire and are replaced each year. Members of the Disciplinary and Appeals Committees will be specifically appointed by Council for each case to ensure that no member of any of the three committees serves on either of the other committees for that case. Further, the Chairmen of the Disciplinary and Appeals Committees will be advised by lawyers retained by the Society.

The Articles of Association of the Society do not differentiate between professional and non-qualified members. An Institutional Member is regarded as being on the same footing as an Affiliate. A professional worker exercises not only the skills which he has learned in his formal education and training, but also mature personal judgement developed from the use of those skills, in the varying situations of his day-to-day working life. The level of a member's professional objectives will be dependent on, amongst other things, his seniority, his position, and his type of work.

Consultants carry special professional obligations. A senior executive in charge of a major computer application or computer project is responsible for the accuracy of the information produced by the installation and for ensuring that those for whom it is prepared are fully aware of its limitations in relation to the purpose for which they intend to use it; he cannot, however, be held responsible if it is used for a purpose of which he is unaware or for which it was not intended. The responsibility of senior systems analysts and programmers is also heightened because their work is so little understood by others and failures can have serious consequences. It must, however, be borne in mind that the more responsibility a member carries, the higher will be the standard expected of him and the more rigorously may the Society's sanctions have to be applied. In the interest of the public, the highest standard will naturally be expected of those in public practice who by the nature of their work accept personal responsibility for what they undertake.

The Society has no legal standing as between a member and his employer, whether an individual or a company. Its remedy lies in giving, where appropriate, fullest support for the stand taken by a member who loses his job, or is in danger of doing so, and of censuring the employer who seeks to place the member in a position which could cause him to violate the ethical code of his profession.

The Society cannot consider a complaint against a member where that member's conduct is the subject of legal proceedings – the Society has no power to take evidence on oath, nor compel the production of documents. In these circumstances a view expressed by any member in his official capacity on behalf of the Society could improperly influence the course of justice. The complaint could only be considered when the legal action is completed, or it is established that no legal proceedings will take place. This does not prevent a member appearing in the courts as an 'expert witness'.

Index